Scientific Analysis of Cultural Heritage Objects

Synthesis Lectures on Engineering, Science, and Technology

Each book in the series is written by a well known expert in the field. Most titles cover subjects such as professional development, education, and study skills, as well as basic introductory undergraduate material and other topics appropriate for a broader and less technical audience. In addition, the series includes several titles written on very specific topics not covered elsewhere in the Synthesis Digital Library.

Scientific Analysis of Cultural Heritage Objects

Michael Wiescher and Khachatur Manukyan

ISBN: 978-3-031-00959-4   paperback
ISBN: 978-3-031-02087-2   ebook
ISBN: 978-3-031-00159-8   hardcover

DOI 10.1007/978-3-031-02087-2

A Publication in the Springer series
SYNTHESIS LECTURES ON ENGINEERING, SCIENCE, AND TECHNOLOGY

Lecture #12
Series ISSN
Print 2690-0300   Electronic 2690-0327

Scientific Analysis of Cultural Heritage Objects

Michael Wiescher and Khachatur Manukyan
University of Notre Dame

SYNTHESIS LECTURES ON ENGINEERING, SCIENCE, AND TECHNOLOGY #12

ABSTRACT

The characterization of cultural heritage objects becomes increasingly important for conservation, restoration, dating, and authentication purposes. The use of scientific methods in archaeometry and conservation science has led to a significant broadening of the field. Scientific analysis of these objects is a challenging task due to their complex composition, artistic and historical values requiring the use of minimally invasive and nondestructive analytical procedures. This textbook summarizes scientific methods that are currently used to characterize objects of cultural heritage and archaeological artifacts.

This book provides a brief description of the structure of matter at the molecular, atomic, and nuclear levels. Furthermore, it discusses the chemical and physical nature of materials from the molecular to the atomic and nuclear level as determined by the principles of quantum mechanics. Important aspects of natural and anthropogenic radioactivity that play a critical role for some of the analytical techniques are also emphasized. The textbook also provides principals and applications of spectroscopic methods for characterization of cultural heritage objects. It describes the technologies with specific examples for utilization of spectroscopic techniques in the characterization of paintings, books, coins, ceramics, and other objects. Analytic approaches that employ isotopes and determination of isotope ratios will be reviewed. General principles of imaging techniques and specific examples for utilization of these methods will also be summarized. In the later part of the book, a number of scientific techniques for the age determination of cultural heritage material and archaeological artifacts will be presented and discussed with specific examples.

KEYWORDS

conservation science, archaeometry, atomic spectroscopy, radiation, archaeological dating

Contents

Preface

The characterization of cultural heritage objects becomes increasingly essential for conservation, restoration, dating, and authentication purposes. The use of scientific methods in archaeometry and conservation science has led to a significant broadening of the field. Scientific analysis of these objects is a challenging task due to their complex composition, artistic and historical values, which require the use of minimally invasive and nondestructive analytical procedures.

This textbook summarizes scientific methods that are currently used to characterize objects of cultural heritage and archaeological artifacts. This book also provides a brief description of the structure of matter at the molecular, atomic, and nuclear levels. Furthermore, it discusses the chemical and physical nature of materials from the molecular to the atomic and nuclear levels as determined by the principles of quantum mechanics. Essential aspects of natural and anthropogenic radioactivity play a critical role in some of the analytical techniques that are emphasized. The textbook also provides principals and applications of spectroscopic methods for the characterization of cultural heritage objects. Technologies are described with specific examples for the utilization of spectroscopic techniques in the characterization of paintings, books, coins, ceramics, and other objects. Analytic approaches that employ isotopes and the determination of isotope ratios are reviewed. General principles of imaging techniques and specific examples for the utilization of these methods are also summarized. In the later part of the book, several scientific techniques for the age determination of cultural heritage materials and archaeological artifacts are presented and discussed with specific examples.

Michael Wiescher and Khachatur Manukyan
Notre Dame, IN
July 2020

Acknowledgments

The authors would like to thank the Faculty Research Support Regular Grant program (FY17 FRSG) of the University of Notre Dame for their support in establishing a cultural heritage research program at the Nuclear Science Laboratory.

Many friends and colleagues have contributed through discussions and information to this book; in particular we like to thank Prof. Walter Kutschera from the University of Vienna for his information and insights over many years of friendship, a friendship that led to the establishment of an AMS program at Notre Dame. Prof. David Ganz from Cambridge University for providing the motivation for moving ahead with the implementation of the PIXE program at Notre Dame. Dr. David Gura, the Curator for Ancient and Medieval Manuscripts and Louis Jordan, the Associate University Librarian, Academic Services and Collections at the Notre Dame library for providing us with many samples from their rich collection for XRF, FTIR, and Raman spectroscopy analysis. The authors also thank Nicole Reifarth and René Reifarth for providing results on the analysis of silver coins using the neutron activation method.

The authors also want to thank Dr. Ed Stech and Dr. Dan Robertson for their support in accelerator operation and for the many participants at the cultural heritage experiments. Prof. Zachary Schultz, now at Ohio State University for his guidance and help in the use of Raman spectroscopy methods at Notre Dame. Last not least we want to thank Prof. Ani Aprahamian for her motivation and support for the project and the many undergraduate students she helped to recruit for the cultural heritage research at Notre Dame.

The authors would specifically like to thank Armenuhi Yeghishyan for her help in the preparation of figures for this book. We also would like to thank Elizabeth Jo for proofreading the text.

Michael Wiescher and Khachatur Manukyan
Notre Dame, IN
July 2020

CHAPTER 1

Science for Cultural Heritage

1.1 INTRODUCTION

Within archaeological artifacts there is a record to which an archaeologist is blind but which a physicist can hope to read.

(M. J. Aitken, Physics Report, 1978)

The goal of the historian, the archaeologist, and the anthropologist is to study and interpret the past. While for many centuries the written or spoken word, the myths and tales saved over time, have provided the primary source for information, the sciences have developed complementary tools to strengthen our knowledge and to improve our ways of understanding and interpreting the daily life of our ancestors. Over the last century, a wide variety of physics methods and techniques have been developed to probe historical, archaeological, anthropological, and art samples. These methods must be suitable for analyzing characteristic signatures that will help to identify a sample and its constituents, or to link the sample to a particular period in history, or a particular artist and his/her tools and techniques. The methods that have been developed typically aim to analyze material structure or content, or determine chemical constituents, or characteristic details of manufacture. These methods must be sensitive on a microscopic level and must distinguish details in the material structure of the artifact well beyond the capabilities of our natural visible senses. As an alternative, methods have been developed that probe the age of historical samples. These "dating" techniques are based on natural clocks that can count time backward through time-dependent physical processes that are inherent to the material being studied.

Both techniques, as well as material analysis and dating, are based on fundamental chemical or physical principles of matter on the molecular, atomic, or nuclear level—the so-called microphysics of matter. This level of analysis provides unique signatures that allow us to look back in time and provide complementary information to the written word of the past. The analytic approach is independent of personal opinion and prejudice, which have so often tainted the historical literature; this approach, therefore, provides in many cases a complementary and frequently a corrective measure to our interpretations and analyses. The scientific analysis of material itself does not provide an evaluation of the historical information. The analysis only answers a question, such as: "the gold content of this Assyrian coin contains X% of a particular metal," or "the flint stone sample from the Neanderthal man's fire is Y years old;" the interpretation and analysis of this information and its implementation in the historical context remain with the historian. While the historian or archaeologist can proceed without knowing the scientific details of the analytical methods and procedures, they should remain aware of the inherent uncertainties of these analytical tools. These uncertainties may affect the interpretation and need to be taken into proper consideration. Uncertainties in the results of the analysis are based on uncertainties at the atomic level—for example, as a result of the Heisenberg Uncertainty Principle or the statistical limitations the scientist has to face when working with minuscule-sized probes and samples. To appreciate these aspects, difficulties, and limitations, the scientific principles of these physics techniques and methods, as well as their limitations, should be understood on a basic level.

Considering the nature and value of historical artifacts, a reliable and efficient physics technique must fulfill two fundamental conditions: first, it must provide a unique signature of the sample, which can be analyzed and linked to the past; and second, it must be non-destructive or must work with only a minuscule sample size. The latter requirement is obvious since no historian would be willing to destroy the sample in order to explore its origin. Again, the requirement for non-destructive techniques points to microscopic analysis methods that can be performed reliably on just a few atoms at the sample level.

There are a multitude of different physics techniques that fulfill the above-listed requirements. The field of applied science methods in art and archaeology has been named *"Archaeometry,"* based on the name of a British scientific journal that has been in publication since 1958. A wide range of these archaeometric techniques are presently being utilized for multiple applications in material analysis and dating.

It is far beyond the scope of this book to provide a complete list of all of the archaeometric methods that have been developed over the last few decades. The purpose of this text is, therefore, to first highlight and clarify the basic principles of the most essential scientific methods and to present in more detail a few of these techniques using specific examples for illustration.

The purpose of material analysis techniques can be classified into two categories. The first category represents information about the originally applied preparation and fabrication techniques of the samples. This provides information about available technologies and the cultural level of the fabrication period. The second category seeks to analyze the origin of the sample. This provides information about migration and trading patterns. Organic material like bone, fat, the leftovers of eaten food, and excrement may provide additional information about the way of life of past generations. Table 1.1 provides a general overview of the goal and target of several material analysis techniques.

Analytical techniques are typically based on molecular, atomic, or nuclear spectroscopy methods, methods that have been identified and developed over the last century as basic science tools for testing with quantum mechanical prediction, regarding the structure of the molecule, the atom, and the internal structure of the nucleus. Atomic techniques are based on the analysis of electromagnetic radiation in the infrared to X-ray range, while the analysis of the chemical structure seeks to identify the molecular components utilizing spectroscopy techniques in the

Table 1.1: Characteristic materials of archaeological study

Artifact		Purpose of Material Analysis	
	Material Identification	Technology Analysis	Location Analysis
Stone			Mineral content
			Tracer elements
Obsidian			Tracer elements
			O, Sr isotopes
Marble			O isotopes
Ceramics		Hardness, element	Tracer elements
		Content, texture	C profile
Glass		Chem. components	
		Silicate structure	
	Analysis of material	Element content	
	Characteristics, chem.		
Metal	Components, molecular	Hardness, alloy structure	
	Structure, elemental	Isotope distribution	
	or isotope components and distribution, texture		
Paint, Tinctures		Chem. structure, element	
		Content, crystal phase	

ultraviolet (UV) to infrared (IR) range. Nuclear spectroscopy techniques focus on the analysis of gamma (γ) as well as alpha (α) and beta (β) particle radiation. All of these methods have been refined with modern high-sensitivity radiation detectors and high-speed data processing techniques. Application for historical and archaeological questions was developed in the middle of the 20th century and has grown rapidly since then. Atomic spectroscopy techniques include atomic absorption and emission spectroscopy, X-ray fluorescence analysis, proton-induced X-ray emission (PIXE), and a multitude of related methods. Molecular spectroscopy relies mainly on Raman and Fourier Transformed Infrared Spectroscopy.

Nuclear spectroscopy methods include: neutron activation techniques, isotope distribution analysis, nuclear resonance analysis, and natural decay measurements. While material analysis techniques are mainly used in elemental and isotopic content interpretation of a sample, dating techniques rely on atomic or isotopic clocks within the sample material.

These methods take advantage of the natural radioactivity inherent in all materials but they also utilize calibration techniques through Dendrochronology and ice core sample analysis. The primary tool is the radiocarbon dating technique. For dating older materials of more than 100,000 years, this approach can be complemented by other techniques such as potassium/argon dating or dating through analysis of the natural decay schemes, such as uranium/thorium dating. An interesting approach, although limited, is the analysis of radiation damage, which may provide information on total radiation exposure and therefore the age. Thermoluminescence is another radiation driven effect, which has been used extensively for age determination, but this method is limited to applications with materials of crystalline structure.

Another aspect of characterizing cultural heritage objects includes imaging at different length scales. Imaging techniques probe the surface or internal layers of objects by recording a signal emitted or reemitted from objects being hit with electromagnetic radiation or electron beams. Depending on the energy and the nature of the materials, the penetration depth of the radiation may vary. This allows for probing of either the surface or the internal structure of objects. Therefore, these methods can be classified into two large groups: surface or bulk imaging techniques. Depending on the scale of observation, these methods can be categorized as atomic, microscopic, mesoscopic, or macroscopic imaging. The most common imaging techniques used in cultural heritage characterization include optical and electron microscopy, radiography and tomography (both X-ray and neutron), reflectometry, and X-ray imaging (or mapping), to name a few.

In the following sections, we will discuss the underlying physics and the principles of the atomic, molecular, and nuclear spectroscopic tools and methods, as they form a basis for discussing the wide variety of spectroscopy applications.

1.2   BASIC TOPICS OF ATOMIC SPECTROSCOPY

"The interior of the atom is as empty as the Universe."

(Philip Lenard, Nobel Prize lecture, 1905)

To reach a sufficient understanding of the method and applicability of spectroscopy with electromagnetic radiation, it is necessary to describe and discuss the underlying physical principles of the origin and nature of electromagnetic radiation. Electromagnetic radiation comes in a broad wavelength range from femto-meter (10–15 m) to meter. Table 1.2 displays the universally accepted prefix system of the basic international standard (SI) unit meter for a length or distance scale, which is traditionally used to describe the electromagnetic wavelength range.

The entire range is called the electromagnetic spectrum as shown in Figure 1.1. The short wavelength range, from 1 fm to 100 nm, corresponds to a fairly high energy transmission, such as γ-radiation, X-rays, and UV radiation that can be damaging to the eye and the human body. The optical range of the electromagnetic spectrum is in the 0.1–1 μm range and corresponds to a small part of the electromagnetic spectrum. Higher wavelengths correspond to the infrared

Table 1.2: Prefix system of the basic international standard (SI) unit meter for a length or distance scale

Power	Prefix	Abbreviation
10^{-15}	femto	f
10^{-12}	pico	p
10^{-9}	nano	n
10^{-6}	micro	μ
10^{-3}	milli	m
10^{-2}	centi	c
10^{-1}	deci	d
10^{3}	kilo	k
10^{6}	mega	M
10^{9}	giga	G
10^{12}	tera	T
10^{15}	peta	P

and radio waves. Infrared radiation extends the optical range to the millimeter range, while radio waves have a wavelength in the meter range. The characteristic wavelengths λ of the electromagnetic spectrum, which are typically used for material analysis purposes, originate on the atomic or molecular level. In the case of infrared spectroscopy, the wavelength is frequently expressed in terms of the k number with $k = \lambda^{-1}$. The wavelength of an electromagnetic wave λ is inverse proportional to the energy associated with the wave.

An important conclusion by Max Planck, in the early days of quantum physics, was that the energy comes quantized in small discrete light quanta or wave packages, with an energy E_v that directly relates to the frequency v or wavelength, of the associated electromagnetic radiation and the speed of light $c = 2.989 \cdot 10^9$ [m/s]. Such a light quantum is often referred to as a photon with an energy of:

$$E_v = h \cdot v = \frac{h \cdot c}{\lambda}. \tag{1.1}$$

The proportional constant h is the so-called Planck constant, one of the fundamental constants in nature, $h = 6.626 \cdot 10^{-34}$ [J $\cdot$ s] $= 4.136 \cdot 10^{-21}$ [MeV $\cdot$ s]. The international standard unit for energy is the Joule [J], and the energy unit for microscopic systems at the atomic or nuclear level is the electron volt [eV]; 1 eV $= 1.602 \cdot 10^{-19}$ J. The small magnitude of the Planck constant represents the dimension of the microphysical, from which the light is being emitted. We therefore concentrate, in the following, on a simplified description of the Bohr model of the atomic nucleus and the origin of its characteristic electromagnetic radiation.

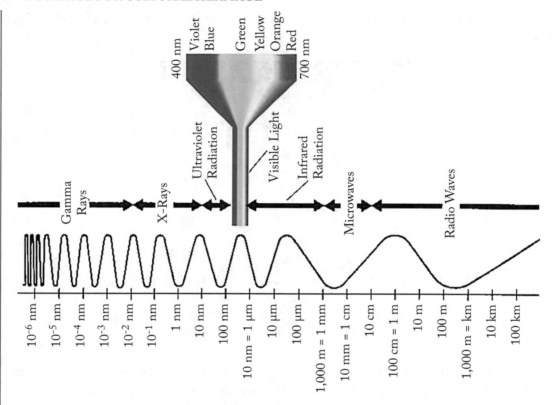

Figure 1.1: Wavelength range for the entire electromagnetic spectrum from gamma rays to radio waves.

The Bohr model of the atomic nucleus, namely the hydrogen atom, was developed by the Danish physicist Niels Bohr in 1913 and is based on the Rutherford alpha particle scattering experiments. These experiments gave evidence that the atom is a void with a mass M being concentrated only in the very center of the atom, giving the atomic nucleus a positive charge $Ze^+ = +Z \cdot 1.6 \cdot 10^{-19}$ C. Z number of electrons are positioned on fixed well-defined orbits around the nucleus. The electrons have a negative charge $e^- = -1.6 \cdot 10^{-19}$ C and the atom as a whole has a neutral charge. The forces that keep the electrons on their orbit result from the attractive electrical (or Coulomb) field between the positively charged nucleus and the negatively charged electrons. While this idea is similar to the well-known concept of planets rotating around the sun on orbits stabilized by the gravitational field, Bohr's microscopic analogy is characterized by one important aspect—the electrons are only allowed on certain fixed orbits with well-defined radii. The radius r_n and the energy level E_n for each electron correspond to a so-called quantum state for that electron, which is defined by a quantum number n. The radius is determined by fundamental constants. For the hydrogen atom, we get the following radius:

$$r_n = \frac{n^2 \cdot \hbar^2}{m_e \cdot k_e \cdot e^2},$$
(1.2)

with n being an integer number $n = 1, 2, 3, \ldots, m_e = 9.105 \cdot 10^{-31}$ [kg] the mass of the electron, e the elementary charge $e = 1.602 \cdot 10^{-19}$ [C], $k_e = 8.9875 \cdot 10^9$ [N $\cdot$ m^2/C^2] the Coulomb constant that defines the strength of the electrical field, and $\hbar = h/2\pi$ the fundamental Planck constant of the microphysics world. The orbit with the smallest possible radius is called the Bohr radius a_0 and corresponds to the quantum number $n = 1$,

$$a_0 = \frac{\hbar^2}{m_e \cdot k_e \cdot e^2} = 0.0529 \text{ [nm]}.$$
(1.3)

We can express the radius of any orbit in the hydrogen atom in terms of its quantum number and the Bohr radius:

$$r_n = n^2 \cdot a_0 = n^2 \cdot 0.0529 \text{ [nm]} \quad n = 1, 2, 3, \ldots.$$
(1.4)

The Bohr radius defines the size of the hydrogen atom in its "ground state," which is the state of the hydrogen atom at the lowest energy configuration.

Each electron orbit corresponds to a different energy level, called the excitation energy. Since the radii are quantized, meaning only fixed radii with integer quantum numbers n are allowed according to Equation (1.4), the possible energy states are quantized, and only certain fixed levels of excitation are allowed for the atom. The energy states can also be expressed in terms of the quantum number n and the Bohr radius a_0

$$E_n = -\frac{k_e \cdot e^2}{2 \cdot a_o} \cdot \left(\frac{1}{n^2}\right) \quad n = 1, 2, 3 \ldots.$$
(1.5)

By inserting numerical values for the constants into the equation, we get the allowed energy levels in units electron volt (1 eV $= 1.602 \cdot 10^{-19}$ J)

$$E_n = -\frac{13.606}{n^2} \text{ [eV]} \quad n = 1, 2, 3, \ldots.$$
(1.6)

These excitation energies correspond to the only allowed quantum states of excitation that the hydrogen atom can reach. If the electron of the hydrogen atom is in the energetically lowest state $n = 1$, the hydrogen atom is in the ground state, $E_0 = -13.606$ eV. If the electron is in the second orbit, $n = 2$, the hydrogen atom is in its first excited state $E_1 = -3.401$ eV, and the electron in the third orbit, $n = 3$, corresponds to the hydrogen atom being in the second excited state, $E_2 = -1.512$ eV, and so on. Atomic physics has developed a special notation for the different orbitals; orbit $n = 1$ is called the K-shell, $n = 2$ the L shell, $n = 3$ the M-shell, and so on. Figure 1.2 shows a level diagram indicating the various excited states the hydrogen

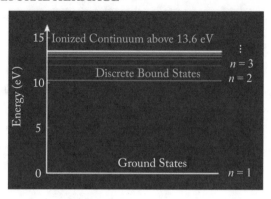

Figure 1.2: Level scheme of the hydrogen atom.

atom can be in. The highest level ($n = \infty$, $E_\infty = 0$) corresponds to the hydrogen ion, where one electron is removed from the atom ($r_\infty = \infty$). This means that the energy necessary for ionizing the hydrogen atom is $E \geq 13.606$ eV, called ionization energy.

To excite an electron from the ground state to an excited level requires a fixed amount of energy, which corresponds to the difference between the excitation energies of the initial ground state $E_i (n = n_i = 1)$ and the excited state $E_f (n = n_f)$. The hydrogen atom absorbs this amount of energy in the form of energy quanta or photons with energy ΔE, which corresponds to the frequency v of the electromagnetic radiation

$$\Delta E = h \cdot v = E_i - E_f = \frac{k_e \cdot e^2}{2 \cdot a_0} \cdot \left(\frac{1}{n_f^2} - \frac{1}{n_i^2} \right). \tag{1.7}$$

The hydrogen atom will subsequently de-excite back to the ground state by emitting a quantized amount of energy, which again corresponds to the energy difference between the initial excited state and the final state. This de-excitation process may occur as a one-step transition but can also take place as a cascade of transitions from the initial excited state to all of the less excited levels at lower energies, as shown in Figure 1.3. Because of the close relationship between energy, frequency, and wavelength of light, as shown in Equation (1.1), the wavelength of the light emitted in such a de-excitation process can also be expressed in terms of the quantum numbers of the initial and final state

$$\frac{1}{\lambda} = \frac{v}{c} = \frac{k_e \cdot e^2}{2 \cdot a_0 \cdot h \cdot c} \cdot \left(\frac{1}{n_f^2} - \frac{1}{n_i^2} \right) = R_H \cdot \left(\frac{1}{n_f^2} - \frac{1}{n_i^2} \right) \tag{1.8}$$

$R_H = 1.09737 \cdot 10^7$ [m^{-1}] is the Rydberg constant of atomic physics. This relation has been empirically determined by Balmer and Rydberg and was viewed as the experimental confirmation of the Bohr model in the early days of quantum theory. Traditionally, the transitions within the hydrogen model are sorted into different series, as indicated in Figure 1.3. All transitions

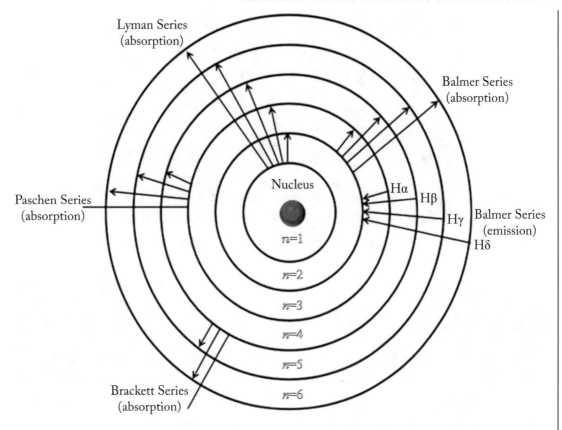

Figure 1.3: Possible electron transitions between the excited states of the hydrogen atom.

from any excited state to the ground state (K-shell) are part of the Lyman series, all transitions to the first excited state (L-shell) belong to the Balmer-series, the transitions to the second excited state (M-shell) belong to the Paschen-series, and the transitions to the fourth excited state (N-shell) are called the Brackett-series. The Lyman-transitions are the most energetic and correspond to ultraviolet light. The Balmer-series is in the range of visible light between violet and red, as shown in Figure 1.4. The Paschen- and Brackett-transitions belong to the infrared, less energetic light.

Because the energy differences and transitions between the various excited states in a hydrogen atom are well defined, the energies or frequencies of the absorbed or emitted photons for the hydrogen atom can be used in spectroscopy to analyze the amount of hydrogen in certain materials.

Figure 1.4 shows the visible part of the hydrogen emission spectrum, the Balmer-series. Other hydrogen lines are observed in the infrared part of the emission spectrum, for example

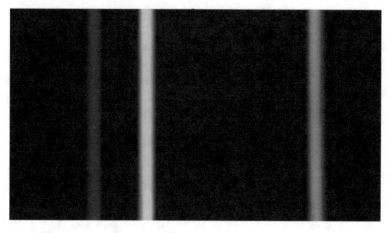

Figure 1.4: Visible part of the hydrogen emission spectrum, the Balmer-series.

Figure 1.5: Visible part of the solar spectrum. The dark lines are the characteristic absorption lines from hydrogen and other elements in the solar photosphere.

in the 21-cm line, which corresponds to the transition between different spin directions of the electron in the K-shell (see discussion of spin quantum number, Equation (1.11)). This radio line is used for the analysis of the hydrogen content of far distant stellar nebulae and is a powerful astronomical tool for the analysis of absorption spectra. The hydrogen abundance in the solar atmosphere, for example, has been analyzed from the solar hydrogen absorption spectrum shown in Figure 1.5. Absorbing the characteristic wavelengths of solar radiation excites the hydrogen atoms in the solar photosphere; the dark lines in the spectrum are the absorption lines and correspond to the characteristic hydrogen energies.

For simplicity reasons, we have confined our discussion to the hydrogen atom; more massive multi-electron atoms with higher Z nuclei require a more complex quantum mechanical treatment that is well beyond the scope of the present text.

The electron orbits are not only distinguished by the main quantum number n but also by the integer orbital quantum number $\ell = 0, 1, 2, \ldots, n - 1$, which associates a certain possible orbital momentum L to an electron in each shell with main quantum number n, with L being

Table 1.3: Atomic shell and sub-shell notations

n	Shell	ℓ	Sub-Shell
1	K	0	S
2	L	1	P
3	M	2	D
4	N	3	F
5	O	4	G
6	P	5	H

calculated in terms of the Planck constant $\hbar$,

$$L = \sqrt{\ell(\ell+1)} \cdot \hbar. \tag{1.9}$$

This means that each orbital n can have electrons in certain orbital momentum configurations ℓ. Electrons in the K-shell, for example, have only orbital momentum $\ell = 0$, which describes a spherical symmetric orbital configuration called s-wave electrons. The electrons in the L-shell can have $\ell = 0$, s-wave configuration or $\ell = 1$, p-wave configuration, the latter corresponding to a more elliptical trajectory of the electron. The notation is summarized in Table 1.3.

The orbital momentum effects the energy of the corresponding quantum level; this has been formulated by the theoretical physicist Arnold Sommerfeld as the Sommerfeld model of the atom. With the energy of a quantum state n, ℓ is expressed in terms of the Sommerfeld fine structure constant α,

$$E_{n,\ell} = E_n \cdot \left(1 + \alpha^2 \cdot \frac{Z^2}{n^2} \cdot \left(\frac{n}{\ell+1} - \frac{3}{4}\right)\right) \quad \text{with} \quad \alpha \approx \frac{1}{137}. \tag{1.10}$$

Another quantum number is the spin quantum number s. This quantum number has only two possibilities $s = \pm 1$ (often described as spin up ↑ or spin down ↓), which can be associated with the two possible rotational directions of the electron around its axis within the framework of the simple model used here. The magnitude of the spin-orbital momentum S is again quantized

$$S = \sqrt{s(s+1)} \cdot \hbar = \frac{\sqrt{3}}{2} \cdot \hbar. \tag{1.11}$$

Other quantum numbers are the magnetic quantum numbers m_ℓ and m_s, which describe the alignment, of the two orbital momenta L and S, that occurs when the electrons couple with an external magnetic field.

Each electron configuration is defined by its unique combination of quantum numbers. The so-called Pauli exclusion principle, formulated by the theoretical physicist Wolfgang Pauli, states that "no two electrons in the same atom can ever be in the same quantum state." Each configuration is well defined by its unique set of quantum numbers, $n, \ell, s, m_\ell,$ and m_s.

As the identification for each possible electron state is a fixed set of quantum numbers, each possible electron configuration corresponds to a unique energy state. These energy states depend not only on the quantum numbers but also on the charge Z of the nucleus and the number Z of the electrons populating the various quantum states. The Coulomb forces involve not only the attractive force between a single negatively charged electron and the positive nucleus but also the deflective forces between the negatively charged electrons in the various shells. This affects, in particular, the energy configuration of the outer shells where the electrons are partly shielded from the positive charge Z of the nucleus by the inner shell electrons. The allowed energies in a multi-electron atom with Z electrons are therefore described in terms of an effective charge Z_{eff}, which not only depends on Z but also on the quantum numbers n, ℓ, s,

$$E_n = -\frac{k_e \cdot e^2}{2 \cdot a_0} \cdot \left(\frac{Z_{eff}^2}{n^2}\right) = -\frac{13.606 \cdot Z_{eff}^2(Z, n, \ell, s)}{n^2} \text{ [eV]}. \tag{1.12}$$

The exact description of the Z and quantum number dependence $Z_{eff}(Z, n, \ell, s)$ is the subject of quantum electrodynamics (QED), which is beyond the scope of this text. Because each element is defined by its number of electrons or nucleus charge Z, each element is characterized by its own unique transitions—by absorption or emission of electrons between the different quantum configurations—and so each atom has its unique spectroscopic signature. While the characteristic signature is fairly simple in the case of the above-discussed hydrogen atom, it can become very complex for higher Z atoms. The study, identification, and theoretical description of higher Z atoms is still an important research subject in the field of atomic physics and in QED.

In relatively simple cases, like transitions of electrons between the K-shell ($n = 1$) and the L-shell ($n = 2$), a direct Z-dependence can be formulated. For the transition to occur, there must be only one electron in the K-shell. Since the electron in the L-shell is shielded by the K-shell electron, it only sees the reduced charge of the nucleus ($Z - 1$); the transition of the L-shell electron back to the K-shell is described by Moseley's law:

$$\Delta E = h \cdot v = E_2 - E_1 = -(Z - 1)^2 \cdot 13.6 \cdot \left(\frac{1}{2^2} - 1\right). \tag{1.13}$$

These are characteristic high energy transitions that emit photons in the X-ray energy range. While the transition between the L-shell and the K-shell in the hydrogen atom corresponds to an energy of $\Delta E = 10.2$ eV, in the case of a tungsten atom with $Z = 74$, the same transition corresponds to significantly higher energy $\Delta E = 54.355$ keV. This is the energy of a photon, which corresponds to the X-ray energy range. The corresponding X-ray wavelength λ for transitions to the K-shell can be expressed, as similar to Equation (1.8), in terms of the Rydberg constant

$$\frac{1}{\lambda} = R_H \cdot (Z - 1) \cdot \left(1 - \frac{1}{n^2}\right). \tag{1.14}$$

For transitions to higher quantum number shells in multi-electron atoms, the internal shielding conditions are more complex and the effective charge is approximated by $Z_{eff} = Z - \sigma$ with σ being a specific constant that depends on the quantum numbers of the associated electron shells

$$\frac{1}{\lambda} = R_H \cdot (Z - \sigma) \cdot \left(\frac{1}{n_f^2} - \frac{1}{n_i^2} \right). \tag{1.15}$$

This equation shows that each transition—between two orbits n_f and n_i in an atom with atomic number Z—corresponds to a well-defined wavelength of the emitted or absorbed photon. This wavelength decreases rapidly with increasing Z; for lead with $Z_{Pb} = 82$ the wavelength for a fixed transition is about ten times smaller than for example with fluorine $Z_F = 9$. This translates into frequencies or photon energies for corresponding transitions that increase with Z. A transition of the K- or L-series in lead would correspond to a ten times higher energy release than the same transition would affect in fluorine. Therefore, X-ray spectroscopy appears as a unique tool for detailed analysis of heavy element content in historic materials.

Often the X-ray spectrum is analyzed in terms of its energies E_x rather than in terms of the X-ray wavelengths λ- and the expressions for the K- and the L-series are then formulated as

$$E_x = (Z - 1)^2 \cdot 13.6 \cdot [\text{keV}] \cdot \left(1 - \frac{1}{n_f^2} \right)$$

$$E_x = (Z - \sigma)^2 \cdot 13.6 \cdot [\text{keV}] \cdot \left(\frac{1}{2^2} - \frac{1}{n_f^2} \right). \tag{1.16}$$

These equations allow a relatively quick and simple calculation, for the energies of the K- and L-transitions, and are mostly used in X-ray based elemental analysis. More detailed information about the characteristic energies, taking fine structure orbital momentum effects into account, are more difficult to calculate. It is, therefore, better to rely on web-based tabulations for these energies, such as: http://www.csrri.iit.edu/mucal.html. Figure 1.6 shows an example of the typical X-ray spectrum for a silver specimen $Z = 47$). Using Equation (1.16), the K_β-line (transition from $n = 2$ to $n = 1$) is expected to show $E_x = 21.58$ keV, and the K_β-line (transition from $n = 3$ to $n = 1$) is expected for $E_x = 25.58$ keV which agrees very well with the observed peak positions.

1.3 INTERACTION OF ELECTROMAGNETIC RADIATION WITH MOLECULES

A voyage to Europe in the summer of 1921 gave me the first opportunity of observing the wonderful blue opalescence of the Mediterranean Sea. It seemed not unlikely that the phenomenon owed its origin to the scattering of sunlight by the molecules of the water.

(C. V. Raman, Nobel Prize lecture, 1930)

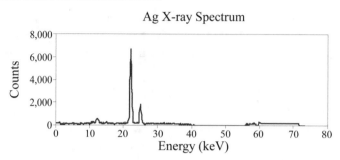

Figure 1.6: X-ray spectrum of a silver specimen. Two peaks are observed for silver and are as-signed as Ag K_α 22.16292 keV and Ag K_β 24.9424 keV lines. Typical Ag L-transitions appear at energies below 3 keV and are not observed in the spectrum.

The combining of atoms into molecules or crystals creates unique energetic states, and interaction with electromagnetic radiation induces different spectra of the transitions between these molecular states. Molecular spectra of materials can be obtained by investigating their unique electronic states, rotations, and vibrations. Electronic energy transitions give rise to absorption or emission in visible and ultraviolet regions of the electromagnetic spectrum. Rotations are collective motions in molecules and can reveal characteristics in the microwave and millimeter spectral regions. Vibrations are relative motions in the molecules or crystal and can be probed by infrared and Raman spectroscopy. The periodic lattice structure in crystalline materials also scatters X-rays, electrons, or neutrons, revealing their structure.

Vibrational spectroscopy includes several techniques, such as IR and Raman spectroscopy; these techniques are prevalent for investigating the structure of cultural heritage objects. Both IR and Raman spectroscopies provide for an investigation of characteristic vibrations, which can be efficiently used for the elucidation of the molecular make-up of materials. Spectroscopies allow for a range of information, from simple identification testing to in-depth quantitative analysis. Test samples can be various parts of a solid object or tiny amounts taken from the object. Raman and IR spectroscopy are complementary techniques although some vibrations may be active in both techniques. Raman spectroscopy is useful for probing symmetric vibrations of non-polar groups, while IR is viable for studying the asymmetric vibrations of polar groups.

Electromagnetic radiation is characterized by wavelength λ, frequency v (the number of vibrations per unit time) and wavenumber k:

$$\Delta E_m = hv = \frac{hc}{\lambda} = hck, \quad \text{or} \quad \frac{\Delta E_m}{hc} = k. \tag{1.17}$$

This equation states that the wavenumber of the absorbed or emitted photon is equal to the change in the molecular energy (ΔE_m) term expressed in cm^{-1}.

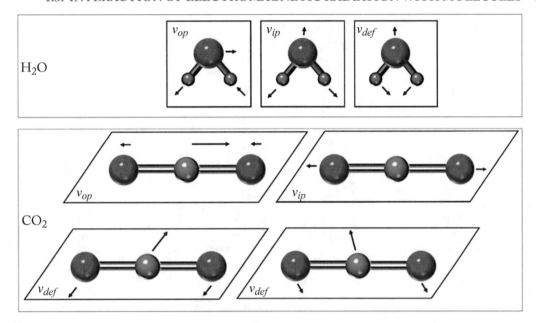

Figure 1.7: Schematic diagrams of H_2O and CO_2 molecular motions.

1.3.1   DEGREES OF FREEDOM AND MOLECULAR MOTION

Let us discuss the model of a molecule, for which mathematical points represent the atomic nuclei without mass. The intermolecular forces holding atoms together are assumed to be as a massless spring, which restores chemical bond length or bond angles to a particular equilibrium value. Each atomic nucleus requires three coordinates (x, y, z) to define its positions in a particular coordinate system. As a result, nuclei will have three independent degrees of freedom of motion in x, y, or z directions. If the molecule consists of N atomic nuclei, there will be a total of 3N degrees of freedom of motion.

The molecular motion that results from characteristic vibrations of molecules is described by the internal degrees of freedom, which result in 3N-6 and 3N-5 rules for vibrations of non-linear and linear molecules, respectively. Figure 1.7 shows the main vibrations for water and carbon dioxide molecules. The water molecule has three atoms and is a nonlinear molecule, which has $3(3) - 6 = 3$ degrees of freedom. These three vibrations include an out-of-phase and an in-phase stretch, and a deformation (bending) vibration, as shown in Figure 1.7. Simple examples of linear molecules include H_2, N_2, and O_2, which all have $3(2) - 5 = 1$ degree of freedom. The only vibration for these simple molecules is stretching vibration. The more complicated CO_2 molecule has $3(3) - 5 = 4$ degrees of freedom and therefore four vibrations. The four vibrations include an out-of-phase, an in-phase stretch, and two perpendicular deformation-bending vibrations (Figure 1.7).

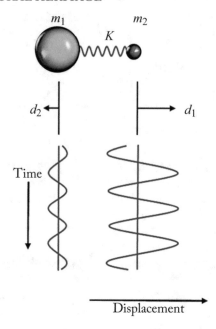

Figure 1.8: The vibration of an HCl molecule showing displacement vs. time as sinusoidal function.

For the molecular vibrations, the displacements of each atom in the molecule change periodically, with the same frequency and passing through equilibrium positions simultaneously. The center of the mass does not move, and the molecule does not rotate. If we consider the motion of atoms in a molecule as being similar to a harmonic oscillator, the displacement of each atom as a function of time can be plotted as a sinusoidal wave. The vibrational amplitudes may differ in either magnitude or direction. Figure 1.8 shows a vibration for the hydrogen chloride (HCl) molecule.

1.3.2   HARMONIC OSCILLATOR

Let us consider a simple harmonic oscillator model to better understand the molecular vibrations that are responsible for the characteristic bands observed in vibrational spectra. Figure 1.8 demonstrates a molecule consisting of two atoms with masses m1 and m2 connected by a spring, which has no mass. The displacement of each atom from equilibrium along the spring axis is d_1 and d_2. The displacement of the two atoms is a function of time, just as a harmonic oscillator varies periodically as a sinusoidal function.

Let us assume that both atoms have the same vibration frequency and both masses pass through their equilibrium positions simultaneously. The observed vibration amplitudes are inversely proportional to the mass of the atoms, which keeps the center of mass unchanged:

$$\frac{d_1}{d_2} = \frac{m_2}{m_1}.$$

The vibrational frequency for a molecule consisting of two atoms will be:

$$v = \frac{1}{2\pi} \sqrt{K \left(\frac{1}{m_1} + \frac{1}{m_2} \right)}, \tag{1.18}$$

where m_1 and m_2 are the masses in grams, K is the force constant in newton per meter (N/m), and v is in cycles per second. This relation can also be expressed by using the reduced mass term:

$$\frac{1}{\mu} = \frac{1}{m_1} + \frac{1}{m_2} \quad \text{or} \quad \mu = \frac{m_1 m_2}{m_1 + m_2}.$$

Wavenumber (k) is more often used in vibrational spectroscopy

$$k = \frac{1}{2\pi c} \sqrt{K \left(\frac{1}{m_1} + \frac{1}{m_2} \right)}, \tag{1.19}$$

where c is the speed of light (cm/s). Equation (1.19) indicates that the frequency of a vibrating system consisting of two atoms is a function of the force constant (bond energy) and the masses of the atoms involved in the vibration.

Table 1.4 shows the experimentally determined characteristic vibration frequencies of some groups. The force constant and consequently the vibration frequencies of a given bond also depend on neighboring atoms or groups. For example, vibration frequency for the C−H bond is ~3,000 cm^{-1}, which decreases to 2,750 cm^{-1} in O=C−H group. Such change is related to the fact that the oxygen atom decreases the electronic density and thus the force constant of the C−H group. Vibration frequency also depends on the mass of the atom. The smaller the atom mass, the higher the frequency as demonstrated by the example of C−H (~3,000 cm^{-1}) and C−D (~2,120 cm^{-1}) stretching frequencies.

Equation (1.19) allows for calculating the approximate values of the force constant for a given bond, based on an experimentally determined wavenumber. For example, the force constant of a single C−C bond is ~500 N/m, while the force constants for C=C and C $\equiv$ C are ~1000 and 1500 N/m, respectively.

1.3.3   QUANTUM HARMONIC OSCILLATOR

The potential energy (E) of the harmonic oscillation of a molecule consisting of two atoms can be written as

$$E = \frac{1}{2} K x^2. \tag{1.20}$$

Table 1.4: Wavenumber regions for some groups

Group	Frequency Region, cm^{-1}
C-Cl	600–800
C-C	~1,000
C=C	~1,650
C≡C	~2,250
C-O	~1,100
C=O	~1,700

A plot of the potential energy E for a simple molecule is a function of the distance (x) between the atoms, which is symmetrical to approximately the equilibrium distance x_0, in which the energy has a minimum value. The force constant K is a measure of the curvature of the potential well near x_0. According to the theory of quantum mechanics, molecules can only exist in quantized energy states. Therefore, molecular vibrational energy only has specific discrete values. Thus, the vibrating molecule can transition from one specific energetic state to another, as shown in the graph of potential energy for the quantum oscillator (Figure 1.9a). These states (or energetic levels) are equidistance, and the energy (E_i) of each state can be determined with the equation

$$E_i = \left(n_i + \frac{1}{2} \right) h\upsilon \qquad n = 1, 2, \ldots, \tag{1.21}$$

where υ represents the vibrational frequency of the oscillator and n is a quantum number, which can only be changed by $\Delta n \pm 1$. The minimum value of the energy $(E_0 = \frac{1}{2}h\upsilon)$ occurs when $n = 0$ at equilibrium distance (x_0). Figure 1.9a shows the potential energy curve for a harmonic oscillator, with the probability functions for the distance x between the atoms in each energy level. According to Heisenberg's uncertainty principle, we cannot determine with certainty the position of the atom during the vibration. Therefore, the position of a particle should be expressed as a probability.

The harmonic oscillator model of molecular vibrations represents an ideal case and in a real molecule, vibrations are limited by the bond dissociation. Non-harmonic oscillator modes provide a better description of molecular vibrations. Figure 1.9b depicts the potential energy curve and levels for a molecule consisting of two atoms, for harmonic and non-harmonic oscillators. The shape of the energy curve is asymmetric, the right-hand portion converging to the dissociation energy of the molecule. In this model, separation between the energy levels decreases at higher vibrational levels until finally the dissociation limit is reached. In the harmonic oscillator model, only transitions to adjacent levels, so-called *fundamental transitions* are allowed ($\Delta n \pm 1$), whereas, for the anharmonic oscillator, $\Delta n \pm 2, 3, \ldots$ transitions or overtones

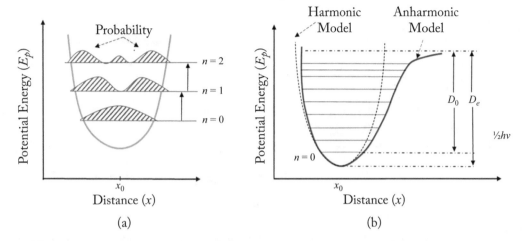

Figure 1.9: The E_p diagram for harmonic (a) and anharmonic (b) oscillator models: transition from $n = 0$ level and D_0 is energy to break the bond.

can also occur. These transitions to higher vibrational levels are much less probable than in the fundamental transitions and are of much weaker intensity.

1.3.4   ABSORPTION PROCESS OF INFRARED LIGHT

The absorption of IR light by a molecule (or groups in crystalline solids) is a quantized process. The molecule absorbs only a selected frequency of IR radiation. The IR radiation absorption corresponds to an energy change of $\sim$10–40 kJ/mol. This energy range encompasses the stretching and bending vibrational frequencies of a typical bond in a molecule. During absorption, the frequencies of IR radiation that match the vibrational frequencies of the molecule are absorbed. The absorbed energy increases the amplitude of the vibrational motion of the specific bond. We should note that not all bonds are capable of absorbing IR radiation, even when the bond motion frequency matches exactly to the IR radiation frequency. This is because the dipole moment of the molecule must be changed in order to absorb the IR photon. The dipole moment (μ) is the measure of the polarity of a chemical bond in a molecule. The dipole moment depends on the magnitude of the charges and their positions and can be derived from partial charges on the atoms. Molecules with only one type of element (such as N_2, O_2) have no dipole moment and thus are IR inactive. HCl, CO_2, or NO molecules have a dipole moment and exhibit IR active vibrations.

The intensity of an absorption process depends on the strength of the process; this process is mainly caused by the change of the electric dipole moment, which has a vibrational transition that is induced by infrared radiation. The probability of a transition with $\Delta n \pm 2$ is ten times smaller than that of $\Delta n \pm 1$.

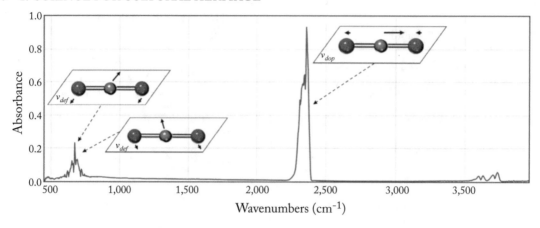

Figure 1.10: IR spectrum of CO_2 molecule. Asymmetric stretch (~ 2360 cm^{-1}) and degenerate banding modes (~ 650 cm^{-1}). Symmetric stretch vibration is IR inactive. Source: NIST Chemistry Webbook (webbook.nist.gov).

In the IR spectrometer, the light source simultaneously emits broad range IR frequencies that interact with the materials under investigation. The Lambert–Beer law determines the absorption (A) of IR light, connects the intensity of incident (I_0) and transmitted (I) lights, and the concentration of the analyzed compound:

$$A = \log\left(\frac{I_0}{I}\right) = \epsilon l c, \tag{1.22}$$

where ϵ is the molar absorptivity of the substance and l is the path length. In a typical IR spectrum, the intensity of absorbed (or transmitted) light is plotted against the wavenumber. Figure 1.10 shows IR spectrum of a CO_2 molecule with asymmetric stretch appearing at ~ 2360 cm^{-1} and two degenerate banding modes with similar vibrational frequency centered at ~ 650 cm^{-1}. The in-phase symmetric stretch of the carbon dioxide molecule shown in Figure 1.7 is not IR active, as the net molecular dipole is not changing during the vibration.

1.3.5   RAMAN SCATTERING PROCESS

IR inactive vibrations, such as the symmetric stretch of CO_2, can be investigated by Raman spectroscopy when the overall polarizability changes during the vibration. On the other hand, the infrared active vibrations of CO_2 (asymmetric stretch and bend) are Raman inactive. In many cases, IR and Raman spectroscopies provide complementary information on the molecular vibrations.

Raman spectroscopy uses visible light to probe molecular vibrations. The process involves shining monochromatic light on the sample. Photons of light interacting with molecules can

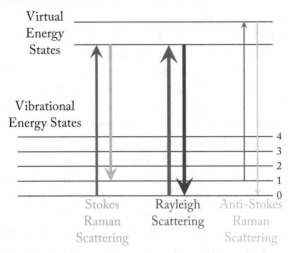

Figure 1.11: Energy level diagram for Rayleigh, Stokes Raman, and anti-Stokes Raman scattering.

scatter by two mechanisms, known as either Rayleigh or Raman scattering. Rayleigh scattering is an elastic process, which does not change the energetic state of the molecule. In this process, most of the photons elastically scatter upon interaction with materials. These scattered photons exhibit the same frequency and wavelength as the incident photons. A small fraction of the scattered photons (0.00001%) has an energy that is different from the incident photon (Raman scattering).

The energy change of molecules interacting with photons by Rayleigh and Raman scatterings is shown in the schematic of Figure 1.11. Upon interaction with light, the molecule is excited, by a photon, from the ground state to a virtual energy state. During relaxation, the molecule returns to the ground state and emits a photon with the same energy as the initial photon (Rayleigh scattering). Or, the scattered light can have smaller frequencies due to the excitation of molecules to a higher vibrational state, the so-called Stokes shift. Alternatively, scattering can occur from an excited state ($n = 1$) relaxing to the ground state ($n = 0$), which is accompanied by the emission of a photon shifted to a higher frequency (the anti-Stokes shift).

A good example for illustrating the Stokes and anti-Stokes shift is the spectra of the carbon tetrachloride (CCl_4) molecule. This molecule exhibits three Raman-active absorptions—at ~ 220, ~ 315, and ~ 460 cm^{-1} shifted from the laser line. Figure 1.12 shows the complete Raman spectrum for CCl_4 with the Stokes and anti-Stokes peaks. The laser experiences an elastic scattering, which reflects the broad peak (Rayleigh scatter) at the middle of the spectrum. The intensity of Rayleigh scattering is significantly higher than the Raman scattering. The anti-Stokes lines (right side of spectra) are less intense and have higher energy than the Stokes lines (left side). This is related to the fact that at ambient temperature most molecules are found in the ground state. Increasing the temperature increase the population of higher energy vibrational states and decreases the population of the ground state. Such distribution of molecules slightly decreases the intensity of the Stokes lines and increases the intensity of the anti-Stokes lines. However, the two sides of the spectrum appear as mirror images of each other with regards to the band positions. The intensity of the Stokes lines depends on the probability of each scattering process and cannot be predicted. The intensity of the anti-Stokes lines decreases with the increase of the Raman shift, due to the lesser number of molecules with higher-energy vibrational levels to produce Raman scattering at that transition.

To enhance Raman scattering one should increase the number of incident photons, in order to produce more molecules in the proper virtual state. High-power monochromatic lasers are preferable as a light source for measuring Raman spectra; these lasers allow accurate measurement of the frequency of the Stokes lines in the spectrum. Another way to enhance measurement is to use array detectors for the simultaneous detection of all scattered radiation.

The intensity of Raman scattered radiation depends on the incident light intensity, number of scattering molecules, frequency of the exciting light, and polarizability of the molecules. Raman signal has several essential parameters for spectroscopy. Since the Raman signal depends on concentration, a quantitation is possible. Raman peak intensity can be increased using shorter wavelength excitation or by increasing the laser power density. Only vibrations that cause a change in polarizability are Raman active. A disadvantage of Raman spectroscopy is that Raman effect is a spontaneous process and the signals are weak. The probability of Raman scattering can be significantly increased by depositing thin layers or nanoparticles of silver or gold onto the surface of the sample. This technique is known assurface enhanced Raman spectroscopy.

1.3.6   SYMMETRY AND VIBRATIONAL SPECTROSCOPY

The symmetry (or the absence of symmetry) defines whether or not Raman or IR spectroscopy can be used to probe the structure of molecules or functional groups. In most cases, Raman spectroscopy can be used to investigate non-polar groups, symmetric, or in-phase vibrations. IR spectroscopy is more efficient for analyzing polar groups, asymmetric, and out-of-phase vibrations. Molecules can be categorized by their symmetry elements, which include *center, planes*, and *axes* of symmetry. A mathematical discipline known as *group theory* uses symmetry elements for predicting the specific vibrations that will be active in IR or Raman spectroscopy. A detailed description of group theory is beyond the scope of this text.

Raman active vibrations of highly symmetrical molecules (such as H_2, CO_2) are IR inactive and vice versa. This is known as the *rule of mutual exclusion* and is generally true for molecules that have a center of symmetry. The rule states that vibration can be active either in the IR or Raman spectra. In such molecules, the vibrations that retain the center of symmetry cannot be studied by IR and may be probed by Raman method. These molecules generate a change in the polarizability during the vibration but there is no change in dipole moment. Conversly, vibrations that do not hold the center of symmetry are Raman inactive. However, these vibrations

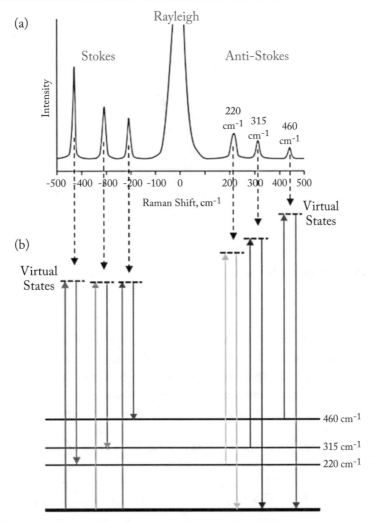

Figure 1.12: Raman spectrum (a) and energy level diagram displaying the origin of Stokes and anti-Stokes lines in the Raman spectrum of CCl_4 (b).

may be IR active as the dipole moment may change. Molecules with no center of symmetry can have some vibrations that are active in both IR and Raman.

Some molecules or functional groups without a center of symmetry may have other symmetry elements that permit them to exhibit vibrations that are active in only one type of vibrational spectroscopy. For example, the in-plane stretches of the NO_3- group in nitrates are Raman active and IR inactive (Figure 1.13). This group has no center of symmetry, but the rapid and simultaneous movement of oxygen atoms results in no change in the dipole moment.

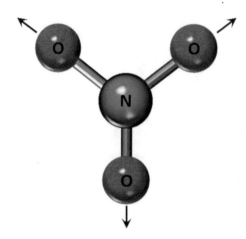

Figure 1.13: In-plane stretching is Raman active, but IR inactive.

Additional symmetry operations such as the plane of symmetry (σ_v), a two-fold rotational axis of symmetry (C_2), and an identity operation (I) define active molecular vibration in Raman or IR spectroscopy. For example, symmetry operations of the water molecule are shown in Figure 1.14. Upon performing symmetry operations, such as rotating the molecule around the C_2 axis, the molecule will have a configuration that is identical to its original configuration before performing such operation.

Figure 1.15 illustrates some relationships of vibrational modes (V_1, V_2, and V_3) of the water molecule and symmetry operations I, C_2, σ_v, and σ_v' using so-called Cartesian displacement vectors. In each combination of a specific vibrational mode and symmetry operation, we assign either ($+1$) or (-1). The use of ($+1$) means that the performed symmetry operation does not result in a change of the original configuration and the resulting new form is symmetrical concerning the specific symmetry operation. In the case of using (-1), all vectors are reversed during the performing of that symmetry operation and the resulting final configuration is said to be anti-symmetrical. Figure 1.15 depicts V_1 and V_2 vibrations that are symmetric to all symmetry operations. The V_3 vibration is anti-symmetric to C_2 and σ_v and symmetric for σ_v' and I symmetry operations. These symmetry considerations allow us to conclude that V_1, V_2, and V_3 vibrations are all active in both Raman and IR spectra. The change in dipole moment for V_1 and V_2 is parallel to the C_2 symmetry axis, and in V_3 vibration, the change is perpendicular to the axis and the σ_v symmetry plane.

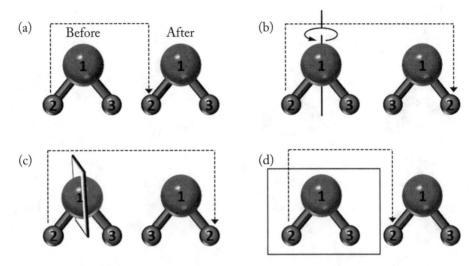

Figure 1.14: Symmetry operations for H_2O molecule in the equilibrium configuration: Identity operation, I with no change (a); rotation by 180°, C_2, two-fold axis of rotation (b); reflection in mirror plane, σ_v (c); and reflection in molecular plane, σ_v' (d).

1.4   BASIC PRINCIPLES OF THE ATOMIC NUCLEUS

Radioactivity is an atomic property of matter and can provide a means of seeking new elements.

(Marie Curie, Nobel Prize lecture 1911)

The atomic nucleus is composed of elementary particles called nucleons. There are two types of nucleons, the protons and the neutrons. Protons carry a positive charge of $e^+ = +1.6 \cdot 10^{-19}$ C and a mass of $m_p = 1.672 \cdot 10^{-27}$ kg; neutrons carry no charge but have a mass slightly larger than the mass of the proton, $m_n = 1.675 \cdot 10^{-27}$ kg. Typically, the mass of these nucleons is expressed in terms of the atomic mass unit u with $1\ u = 1/12 \cdot m(^{12}C)$. The mass of the ^{12}C atom has been measured with high accuracy, $m(^{12}C) = 1.99 \cdot 10^{-26}$ kg, therefore $1\ u = 1.66 \cdot 10^{-27}$ kg. Each nucleus is characterized by its number of protons, Z, and its number of neutrons N. The nucleus has therefore a total positive charge of $Q^+ = Z \cdot e^+ = Z \cdot 1.6 \cdot 10^{-19}$ C. The atomic number Z of protons corresponds to the number of electrons in the atomic shells, therefore the atom itself remains neutral. The total number of nucleons inside the nucleus is the mass number $A = Z + N$. These three numbers uniquely determine the nucleus; it is therefore convenient to use the symbol $^A_Z X_N$ where X represents the chemical symbol for the element. For example, $^{12}_6C_6$ represents the carbon atom with 6 protons and 6 neutrons and the mass number 12. The nuclei of atoms of a particular element with an atomic number Z, can have a different number of neutrons, as with $^{208}_{82}Pb_{126}$ which represents lead with 82 protons, 126 neutrons, and

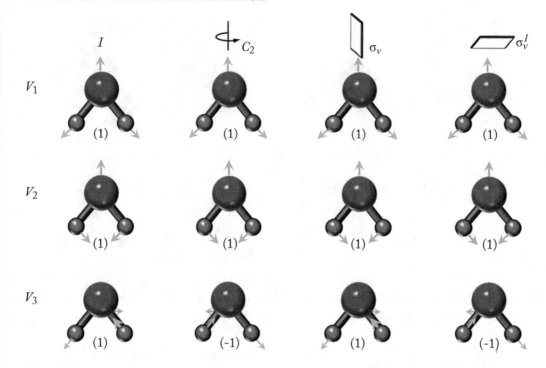

Figure 1.15: Symmetry operations and three fundamental vibration modes (V_1, V_2, and V_3) of H_2O molecule.

the mass number of $A = 208$. The chemical properties remain, however, since these properties are only determined by the number of electrons Z in the outer atomic shell.

Nuclei with a constant atomic number Z but different neutron number N and mass number A are called *isotopes*; nuclei with a constant number of neutrons N but varying atomic number Z and mass number A are called *isotones*, while nuclei with constant mass number A but different atomic number Z and neutron number N are called *isobars*. For example, the element oxygen ($Z = 8$) has three stable isotopes $^{16}O_8$, $^{17}O_9$, $^{18}O_{10}$ with neutron numbers $N = 6, 7, 8$ and mass numbers $A = 16, 17, 18$, respectively. There are also unstable radioactive oxygen isotopes ranging $^{14}O_8$ to $^{26}O_8$. There are three know isobars for mass $A = 18$, $^{18}_9F_9$, $^{18}_8O_{10}$, $^{18}_7N_{11}$ with different atomic number Z and neutron number N. The experimentally known isotones corresponding to $^{18}O_{10}$ with $N = 10$ are $^{16}_6C_{10}$, $^{17}_7N_{10}$, $^{18}_8O_{10}$, $^{19}_9F_{10}$, $^{20}_{10}Ne_{10}$, $^{21}_{11}Na_{10}$, $^{22}_{12}Mg_{10}$, $^{23}_{13}Al_{10}$, $^{24}_{14}Si_{10}$. Such a scheme can be developed for each mass range of the nuclei. This is demonstrated in the nuclear chart, which displays all nuclei as a function of their atomic number Z and neutron number N. Figure 1.16 shows the nuclear chart for all known nuclei. The nuclei marked in black are the stable nuclei; the ones marked in color represent unstable radioactive

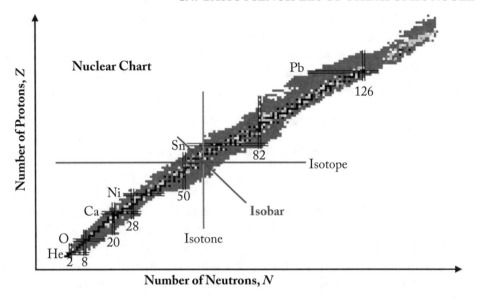

Figure 1.16: Atomic number Z vs. relative to neutron number N for all known nuclei. The stable nuclei are marked in black, radioactive nuclei are marked in color, depending on the kind of radioactive decay.

nuclei, which we will discuss in the following section. Also marked are the position of isotopes (Z = const), isotones (N = const), and isobars (A = constant) within the nuclear chart.

While the Coulomb force binds the electrons in the atoms to the nucleus, the nucleons are bound within the nucleus by the strong force. The strong force is a short-range force, which only extends over a distance of a few femtometer (fm), sometimes called fermi. The strong force exceeds the strength of the deflective Coulomb force, between the positively charged protons, by several orders of magnitude. The range of the strong force determines the size of the nucleus. The radius r of the nucleus can be expressed in terms of its mass number A and a constant $r_0 \approx 1.25$ fm (roughly the size of a nucleon), which approximates the range of the strong force

$$r = r_0 \cdot A^{1/3}. \tag{1.23}$$

Using the average mass of the proton and neutron to be $m_{p,n} = 1.67 \cdot 10^{-27}$ kg, allows for quick calculation of the density of the nucleus ρ_n with a spherical volume of

$$V_n = \frac{4}{3} \cdot \pi \cdot r^3, \qquad \rho_n = \frac{A \cdot m_{p,n}}{\frac{4}{3} \cdot \pi \cdot r^3} = \frac{3 \cdot m_{p,n}}{4 \cdot \pi \cdot r_0^3} = 2.3 \cdot 10^{17} \text{ [kg/m}^3\text{]}. \tag{1.24}$$

This is an enormous density compared to the density of water $\rho_w = 1103$ kg/m^3. The energy, which keeps the nucleons together, is called the binding energy. The binding energy is the amount of energy that is necessary to break the nucleus apart into its single proton and neutron constituents. On the other hand, the binding energy is the energy released if Z protons and N neutrons fusion to a nucleus of mass number A. The binding energy E_B is calculated as the difference between the mass of the nucleus M_A and the total mass of its single proton and neutron constituents, $N \cdot m_n + Z \cdot m_p$

$$E_B = \left(Z \cdot m_p + N \cdot m_n - M_A\right) \cdot c^2. \tag{1.25}$$

With the masses in atomic mass units u, the binding energy (in units MeV) thus translates to

$$E_B = \left(Z \cdot m_p + N \cdot m_n - M_A\right) \cdot 931.494 \; [\text{MeV/u}]. \tag{1.26}$$

Treating the energy conditions for electrons in the central Coulomb field of the nucleus is rather complex but nevertheless can still be calculated with enormous accuracy in terms of QED. However, the energy conditions inside the nucleus itself can be described only approximately. Different model approaches have been developed over the last five decades to describe various aspects of the nucleus and its structure. The liquid drop model is able to describe the general aspects of the binding energy and the general characteristics of nuclear fission processes. The liquid drop model approximates (Figure 1.17) the binding energy E_b by the so-called semi-empirical binding energy formula or Weizsäcker mass formula, as function of mass number A and number of protons Z

$$E_b = a_1 \cdot A - a_2 \cdot A^{2/3} - a_3 \cdot \frac{Z(Z-1)}{A^{1/3}} - a_4 \cdot \frac{N(-Z)^2}{A} \tag{1.27}$$

with the empirical parameters $a_1 = 15.7$ MeV, $a_2 = 17.8$ MeV, $a_3 = 0.71$ MeV, and $a_4 = 23.6$ MeV.

The nuclear forces binding all nucleons together are expressed in terms of a volume effect proportional to A, reduced by a surface effect—accounting for the reduced binding of nucleons at the surface of the nucleus. The binding energy is further reduced by the Coulomb repulsion between the positively charged protons in the nucleus and a term that relates to the large asymmetry between neutron numbers and proton numbers in the nucleus. Modern, more sophisticated mass formulae also include additional microscopic terms, which are beyond the scope of this book. The liquid drop model describes the overall characteristics of a nucleus in terms of its mass and stability. In particular, it is of enormous relevance in describing the phenomenon of radioactivity and the associated possible decay mechanisms for an unstable configuration of nucleons in a nucleus relative to a more stable one; this decay process results in an energetically more favorable nucleon configuration, which corresponds to a lower total mass.

While the liquid drop model semi-classically describes the nucleus in terms of the collective behavior of all nucleons, alternative approaches resulted in the single particle model, which describes the behavior of a single nucleon within the strong interaction potential of all the other nucleons in the nucleus. This model resembles the atomic model, which describes the behavior of a single electron in the atomic shell in terms of its Coulomb interaction with the atomic

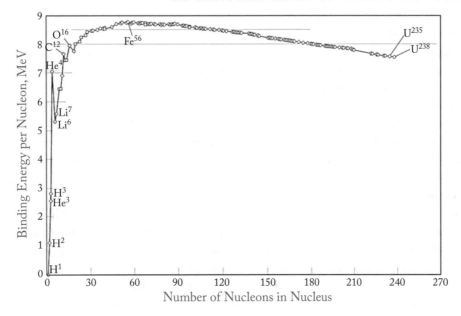

Figure 1.17: The binding energy (E_b) of the nucleus as a function of mass number A. The solid line represents the prediction of the Weizsäcker formula, while the dots show the experimentally observed masses of nuclei plotted as function of mass number A.

nucleus. The single particle model is, therefore, often called the nuclear shell model. In nuclear potential, the single nucleon can only exist in quantized energy states (the nuclear shells), which are described by analogy to the atomic shell configuration with a well-defined set of quantum numbers, n, ℓ, s, m_ℓ, and m_s. Here, the Pauli principle prevails—no quantum state can be populated twice, meaning in practical terms that nucleons cannot collide with each other within the compound of the nucleus. The lowest energetically allowed quantum state is called the ground state of the nucleus; all quantum states of higher energy are called excited states of the nucleus. The shell model describes, similar to its atomic counterpart, the energetics of single nucleon transitions between the allowed quantum states, and so the associated absorption and emission of energy as photons or as kinetic energy of particles. Besides the single-particle quantum state configurations, there exist other more collective quantum state configurations that correspond to different modes of rotation, vibration, and deformation of the nucleus, which will not be discussed here in further detail.

One of the most fascinating phenomena in nuclear physics is the radioactive decay of nuclei. This phenomenon has developed as a major component in a wide variety of industrial, technological, geological, and medical applications. Radioactive decay has also emerged as a unique tool for the analysis and dating of historical artifacts because of its unique signature and the inherent nuclear clock, which controls the time scale of the decay process. In general, the radioactive decay phenomenon can be described as an internal transition of an unstable nucleon configuration into an energetically more favorable, stable configuration. The initial unstable nucleus is called the "mother" nucleus and the decay product is called the "daughter" nucleus.

The probability of radioactive decay depends on the initial and final nucleon configuration of the nuclei and on the associated energy configuration and energy release in the radioactive decay. The decay probability directly correlates with the time scale for the decay process. High decay probability results in rapid decay on a very short time scale, low decay probability is associated with very long time scales. These transitions are mathematically described within the framework of the quantum theory of multi-nucleon systems, which is beyond the scope of the present book.

There are three different types of radioactive decay, α- or more generally particle decay, β-decay, and γ-decay. Figure 1.18 shows a schematic picture for each of the discussed decay processes. In the case of particle decay, the radioactive nucleus decays to a more stable configuration by emitting single nucleons or nucleon configurations. This is known as proton (p) or neutron (n) decay for the emission of single protons or neutrons, or as alpha (α) decay for the emission of a helium nucleus with two protons and two neutrons. The most frequently occurring natural particle decay process is the α-decay. Here are examples of possible particle decay processes involving light, unstable isotopes

$$
\begin{aligned}
\textit{proton decay}: \quad & {}^{5}_{3}\text{Li}_2 \Rightarrow {}^{4}_{2}\text{He}_2 + p \\
\textit{neutron decay}: \quad & {}^{5}_{2}\text{He}_4 \Rightarrow {}^{4}_{2}\text{He}_2 + n \\
\textit{alpha decay}: \quad & {}^{8}_{4}\text{Be}_4 \Rightarrow {}^{4}_{2}\text{He}_2 + {}^{4}_{2}\text{He}.
\end{aligned}
$$

Particle decay is possible if the total mass of the decay products is smaller than the mass of the initial nucleus; according to Einstein's famous equation, the mass difference Δm is proportional to the value $Q = \Delta m \cdot c^2$ of the decay process with c being the speed of light. The released energy Q will be carried away as kinetic energy of the decay products. Particle decay is typically driven by the strong interaction, which keeps the nucleus confined to its small size. Proton and neutron decay only take place for the so-called drip-line nuclei; this has no application for specific analysis studies and will not be discussed further.

The case of α-decay is quite different; α-decay mainly occurs for very massive elements above $A = 200$ (Figure 1.18). Likelihood of α decay mainly depends on the energy difference between the initial nucleus $A = Z + N$ and the final configuration $A - 4 = (Z - 2) + (N - 2) + \alpha$. The decay will occur if the daughter configuration resembles an energetically more favorable state. The energy difference between the two configurations corresponds to the Q-value of the decay, which corresponds to the total kinetic energy of the reaction products because of the mass difference, primarily the kinetic energy of the α particle. The Q-value for the α decay

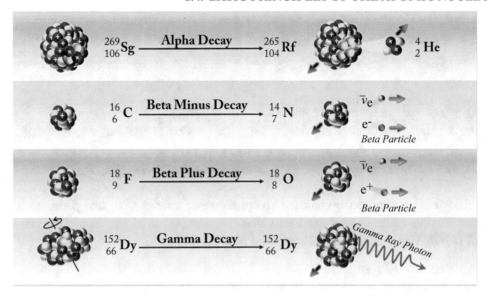

Figure 1.18: Typical decay processes of radioactive nuclei, by α-, β-, and γ-emission.

of a nucleus with mass number A and proton number Z can be calculated by

$$Q_\alpha = M_{A,Z} \cdot c^2 - \left(M_{(A-4,Z-2)} \cdot c^2 + m_\alpha \cdot c^2 \right)$$
$$= E_{kin} \left(M_{(A-4,Z-2)} \right) + E_{kin} \left(m_\alpha \right) . \tag{1.28}$$

Similar expressions can be formulated for proton and neutron decay—or any other form of particle decay. The probability for the decay depends, however, on the probability that the α particle has sufficient energy to penetrate the deflective Coulomb potential of the nucleus, the so-called penetrability. The higher the α particle's kinetic energy, the higher the penetrability and therefore the decay probability.

   Particularly interesting are the natural α-decay chains, which originate by α-decay of long-lived uranium, thorium, and neptunium isotopes. Stellar nucleosynthesis processes have produced these isotopes prior to the formation of the earth and the solar system. The slow decay of these long-lived isotopes initiates a sequence of α- and β-decay processes until the decay chain ends at stable lead isotopes (Figure 1.19). The decay time scale and the impact of the decay processes on the final isotopic abundance distribution in the artifact have emerged as a major tool for archaeological dating and provenance studies.

   The second type of radioactive decay mechanism, the β-decay corresponds to the emission of an electron (β^-) or positron (β^+) particle from the nucleus (Figure 1.18). Electron emission causes transformation of a neutron within the nucleus into a proton, while positron decay converts a proton into a neutron. These decay mechanisms correspond to the conversion of isobars, since the mass number $A = N + Z$ remains constant while N and Z are changed. Typical ex-

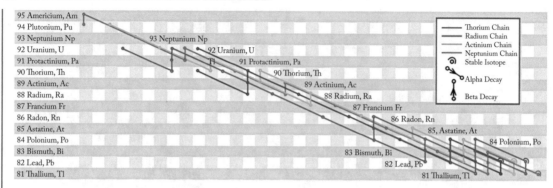

Figure 1.19: Four natural decay chains, each consisting of a series of α- and β-decays initiated by the radioactive decay of long-lived actinides: thorium ($4n$, in blue), neptunium ($4n + 1$, in purple), radium ($4n + 2$, in red), and actinium ($4n + 3$, in green).

amples are the decay of radioactive ^{18}N and ^{18}F isotopes to a stable ^{18}O isotope

$$\beta^- \ \text{decay:} \ {}^{18}_{7}\text{N}_{11} \Rightarrow {}^{18}_{8}\text{O}_{10} + \beta^- + \bar{v}$$
$$\beta^+ \ \text{decay:} \ {}^{18}_{9}\text{F}_{9} \Rightarrow {}^{18}_{8}\text{O}_{10} + \beta^+ + v.$$

The β-decay is controlled by the weak interaction, which determines the probability for internal reconfiguration of the proton and neutron numbers. The β^+-decay reflects the fact that a nucleus has an insufficient number of neutrons to maintain stability, due to the deflecting Coulomb forces of the protons in the nucleus. Therefore, a proton is converted into a neutron by emission of a positron β^+ and a small neutral particle, the neutrino v. On the other hand, β^- unstable nuclei have too many neutrons compared to protons in the nucleus and seek to obtain a more stable configuration by converting a neutron to a proton through emission of an electron β^- and an anti-neutrino $\bar{v}$. The β particles and the neutrinos v are emitted with an energy, which corresponds to the energy difference between the mass of the unstable nucleus and the final stable decay products. The total released decay energy in an electron (β^-) decay is expressed in terms of the Q_β-value

$$Q_{\beta^-} = M_{A,Z} \cdot c^2 - M_{A,Z+1} \cdot c^2 \tag{1.29}$$

and is distributed in the kinetic energy of the emitted electron and anti-neutrino. The Q_β-value for the positron decay can be expressed by a similar equation

$$Q_{\beta^+} = M_{A,Z} \cdot c^2 - M_{A,Z-1} \cdot c^2 - 2m_\beta \cdot c^2 \tag{1.30}$$

with the additional term $2m_\beta \cdot c^2 = 1.022$ MeV, which corresponds to the annihilation energy of the positron, which has to be provided for by the decay mechanism. If the energy difference between the unstable proton-rich nucleus and the decay product is less than the annihilation

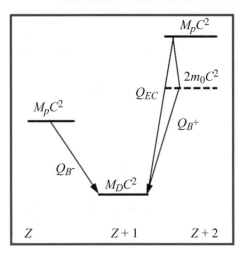

Figure 1.20: Energy relations for the three different kinds of beta decay processes, β^--, β^+-, and electron capture EC. The Q-value for the β^+- decay is indicated, limited by its characteristic energy threshold of $2 \cdot mc^2 = 1022$ keV.

energy of 1.022 MeV, the positron decay is energetically not possible. However, there is an alternative possibility—electron capture into the nucleus to convert one of the protons inside the nucleus to a neutron. The electron is typically captured from the inner K-shell of the atom, which subsequently will be refilled by atomic electron transitions accompanied by characteristic X-ray emission. Neglecting the atomic binding energy of the K-electron, the Q_β-value for the electron capture is expressed by

$$Q_{e-} = M_{A,Z} \cdot c^2 - M_{A,Z-1} \cdot c^2 \tag{1.31}$$

indicating that electron capture is already energetically possible where positron decay is still forbidden. A well-known case is the electron capture decay of ^{7}Be with a $Q_{e-} = 0.862$ MeV, which is clearly below the required minimum value of 1.022 MeV for positron decay.

Figure 1.20 illustrates the energy relations of the various kinds of beta decay processes. The Q-values are indicated with $2 \cdot mc^2 = 1022$ keV as a threshold for the β^+ decay. If the Q_β-value is large, the β-decay often does not directly populate the ground state of the daughter nucleus but, rather, occupies the higher excited states in the daughter nucleus configuration, which subsequently decay by emission of γ-radiation to the ground state. Since the energy difference between excited states and ground state is characteristic for each nucleus, this so-called β-delayed γ-radiation can be used as a characteristic signature for the decay of a particular nucleus.

The γ-decay is, as just mentioned, the decay of excited nuclear states with excitation energy E_x into less excited states or the ground state E_{gs} of a nucleus by emitting high energy electromagnetic radiation. This is very similar to the model of atomic transitions emitting electromagnetic radiation in the infrared to X-ray wavelength range. The photons that are emitted because of nuclear de-excitation processes are much higher in energy, $E_\gamma = h \cdot v = E_x - E_{gs}$, ranging from several keV (similar to X-rays) to several MeV. For example, in ^{12}C the transition of the first excited state to the ground state emits γ-radiation with a characteristic energy of $E_\gamma = 4.4\,\text{MeV}$

$$^{12}_{6}\text{C}^* \Rightarrow {}^{12}_{6}\text{C}_{gs} + \gamma. \tag{1.32}$$

The γ-decay only occurs after the nucleus has been transferred to a higher excited state; this can happen by absorption processes similar to the atom analogy, or by feeding the state through β-decay processes or nuclear reaction processes. The γ-decay processes are typically very fast, compared to α-decay and β-decay processes; they therefore are less useful for actual dating applications and are mainly used as characteristic signatures for identifying reaction or decay products.

1.5   THE LAWS OF RADIOACTIVE DECAY

The activity of Ur X and Th X decays according to an exponential law with time. This, we shall see later, is the general law of decay of activity in any type of active matter.

(Ernest Rutherford, Radioactivity, 1904)

The time dependence of all decay processes can be expressed in the general terms of simple exponential behavior, independent of the actual nature of the decay mechanism. The number of radioactive decays from a radioactive sample within a given time is expressed in terms of the activity $A(t)$ at time t

$$A(t) = -\frac{dN}{dt} = \lambda \cdot N(t). \tag{1.33}$$

The activity is directly proportional to the number of radioactive nuclei in the sample $N(t)$; the constant λ is the decay constant and corresponds directly to the quantum mechanical probability that the nucleus changes from its initial "mother" configuration of proton and neutrons into a different "daughter" configuration. The minus sign in the equation indicates that the number of radioactive nuclei $N(t)$ decreases with time. This equation allows for calculating the number of radioactive nuclei in the sample at any time t after their original production ($t = 0$) of the initial amount N_0

$$N(t) = N(t = 0) \cdot e^{-\lambda \cdot t} = N_0 \cdot e^{-\lambda \cdot t}. \tag{1.34}$$

This is the radioactive decay law, which expresses the exponential decay of every kind of radioactive nucleus with time. Similarly, the associated radioactive activity $A(t)$ can be expressed by the decay law in terms of the initial activity A_0

$$A(t) = A_0 \cdot e^{-\lambda \cdot t}. \tag{1.35}$$

Since the decay constant λ defines the time scale of the decay, it has conventionally been used to define the so-called lifetime τ of the radioactive nucleus

$$\tau = \frac{1}{\lambda}. \tag{1.36}$$

Since the decay process follows an exponential law, lifetime does not mean that after time τ the radioactive sample is gone but, rather, it reflects a statistical expectation value for the decay process to occur. After τ time, the number of radioactive nuclei is reduced to $\approx 66\%$ of its original value since

$$N(\tau) = N_0 \cdot e^{-\lambda \cdot \tau} = N_0 \cdot e^{-1} \approx 0.66 \cdot N_0. \tag{1.37}$$

Instead of the sometimes misleading expression "lifetime," "half-life" $T_{1/2}$ of a radioactive nucleus is used. The half-life corresponds to the time period in which a number of radioactive nuclei N_0 (or the corresponding activity A_0) have decayed to 50% of the original value,

$$N\left(t_{1/2}\right) = N_0 \cdot e^{-\lambda \cdot t_{1/2}} = \frac{1}{2} \cdot N_0. \tag{1.38}$$

Therefore, the decay constant λ can be expressed in terms of the half-life $T_{1/2}$

$$\frac{1}{2} = e^{-\lambda \cdot t_{1/2}} \quad \Rightarrow \quad \ln 2 = \lambda \cdot t_{1/2} \quad \Rightarrow \quad t_{1/2} = \frac{\ln 2}{\lambda}. \tag{1.39}$$

Figure 1.21 shows a typical radioactive decay curve for the radioactive isotope radium ^{226}Ra, which is a α-emitter with a half-life of $T_{1/2} = 1620$ years (see also Example 1.1). The figure shows the exponential decay of the initial activity (or the number of radioactive isotopes) with time. It indicates that after each half-life, the initial amount (or corresponding activity) of ^{226}Ra is reduced to half of its previous value.

Each radioactive isotope has a characteristic half-life. Experimental determination of the half-life, together with the nature of the decay (α-, β-, γ-decay), can serve as a truly unique signature for identification of the isotope. The half-lives of radioactive isotopes range from fractions of a second to billions of years and can be found tabulated in nuclear data tables. For example, the radioactive ^{11}C isotope (see Example 1.1) with 6 protons and 5 neutrons has a half-life of $T_{1/2} = 20.4$ m, ^{12}C and ^{13}C are stable, but ^{14}C has a half-life of $T_{1/2} = 5730$ year, while ^{15}C with 6 protons and 9 neutrons again has a short half-life of $T_{1/2} = 2.45$ s. This demonstrates the impact of the neutron number on the overall stability of a nucleus.

The activity $A(t)$ of a radioactive sample is directly proportional to the number of radioactive isotopes in the sample and inversely proportional to its half-life $T_{1/2}$

$$A(t) = \lambda \cdot N(t) = N(t) \cdot \frac{\ln 2}{T_{1/2}}. \tag{1.40}$$

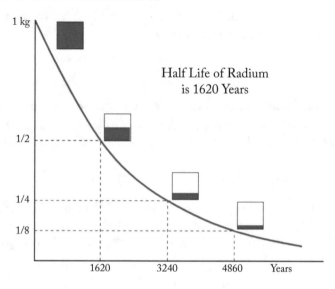

Figure 1.21: Radioactive decay curve for ^{226}Ra.

Example 1.1   Consider the two long-lived radioisotopes ^{226}Ra and ^{14}C with a half-life of $T_{1/2} = 1600$ y and 5730 y, respectively. For a 10,000-year-old artifact, which originally contained 10^6 ^{226}Ra isotopes and $5 \cdot 10^4$ ^{14}C isotopes, how many isotopes of each kind are still in the sample?

$N_{226_{Ra}}(10000\,[\text{y}]) = 10^6 \cdot e^{-\lambda \cdot 10000\,[\text{y}]}$ $N_{14_C}(10000\,[\text{y}]) = 5 \cdot 10^4 \cdot e^{-\lambda \cdot 10000\,[\text{y}]}$

$\lambda_{226_{Ra}} = \ln 2/1600\,[\text{y}] = 4.33 \cdot 10^{-4}\,[\text{y}^{-1}]$ $\lambda_{14_C} = \ln 2/5730\,[\text{y}] = 1.21 \cdot 10^{-4}\,[\text{y}^{-1}]$

$N_{226_{Ra}}(10000\,[\text{y}]) = 13139$ $N_{14_C}(10000\,[\text{y}]) = 14910.$

This means that material containing a large number of radioactive isotopes with short half-lives shows much higher activity than a comparable number of radioactive isotopes with long half-lives. If the half-life is known, a measurement of the activity directly yields the number of radioactive isotopes in the sample. The activity is expressed in terms of the unit Becquerel (Bq), named after the French physicist Henri Becquerel who first discovered the phenomenon of radioactivity. An activity of 1 Bq corresponds, by definition, to one radioactive decay process per second. An older unit for the activity is the Curie (Ci)—named after the Polish-French chemist Marie Curie—one Curie is defined as the activity of 1 g radium, which emits $3.7 \cdot 10^{10}$ α particles per second. This unit is still in use today and therefore often requires conversion between the two units: 1 Ci = 3.710^{10} Bq. With the decay of a radioactive particle such as ^{14}C,

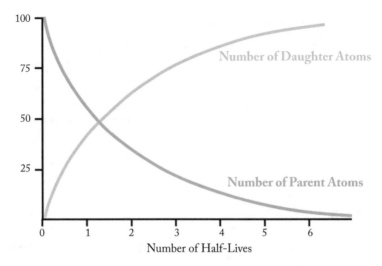

Figure 1.22: Generic decay of a mother isotope and formation of a stable daughter isotope as a function of half-life. After a half-life period, 50% of the radioactive isotope has decayed and now forms the stable daughter isotope.

its initial abundance N_0 declines, building up the abundance of the decay product ^{15}N as shown in Figure 1.22.

Example 1.2   A small 25-g piece of wood, which originally contained $N_0 = 1.6 \cdot 10^{12}$ atoms of ^{14}C, still shows an activity of $A(t) = 15,000$ Bq. Determine the age of the wood sample.

$$t = \frac{5730 \text{ y}}{\ln 2} \cdot \ln \left(\frac{\ln 2}{5730 \text{ y}} \cdot \frac{1.6 \cdot 10^{12}}{15,000 \text{ Bq}} \right)$$

$$= 8267 \text{ y} \cdot \ln \left(1.21 \cdot 10^{-4} \text{ y}^{-1} \cdot \frac{1.6 \cdot 10^{12}}{4.76 \cdot 10^{-4} \text{ y}^{-1}} \right) = 220,985 \text{ y}.$$

In many cases, the radioactive nucleus does not decay to a final stable daughter nucleus but, instead, decays to another radioactive isotope, which in turn decays further as shown in Figure 1.23. In this way, long chains of decay processes can occur until finally a stable isotope is reached. Each decay process along this chain has its characteristic decay mechanism and decay time (or half-life). Consider a decay chain initiated by the radioactive decay of isotope

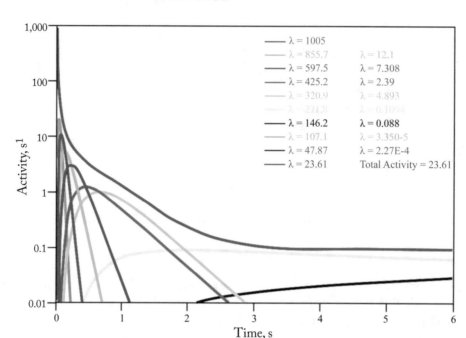

Figure 1.23: Solution of the Bateman equation for a β-decay chain toward stability for a sequence of $A = 156$ radioactive isotopes with different decay constants λ_i. The activity on logarithmic scale is shown as a function of time for all members of the decay chain. The original material will β-decay with this characteristic exponential behavior building the abundance and activity of the first daughter, to be followed by the build-up and decay of the second daughter, and so on until a final stable isotope is reached. The total activity is the sum of the activities of all isotopes along the chain at every point of time.

$A \Rightarrow B \Rightarrow C \Rightarrow \ldots \Rightarrow X$ with isotope X being stable. The characteristic decay constants are $\lambda_A, \lambda_B, \lambda_C, \ldots$; the decay of A feeds isotope B, its decay feeds isotope C, and so on. Equilibrium between the different decay and formation processes evolves; within that equilibrium, the decay and production of the various isotopes along the chain depend on the decay constants and can be expressed in terms of the so-called Bateman equation

$$N_A(t) = N_A(t = 0) \cdot e^{-\lambda_A \cdot t}$$

$$N_B(t) = \frac{\lambda_A}{\lambda_B - \lambda_A} \cdot N_A(t = 0) \cdot \left(e^{-\lambda_A \cdot t} - e^{-\lambda_B \cdot t} \right)$$

$$N_C(t) = \frac{\lambda_A \cdot \lambda_B}{\lambda_B - \lambda_A} \cdot N_A(t = 0) \cdot \left(\frac{1 - e^{-\lambda_A \cdot t}}{\lambda_A} - \frac{1 - e^{-\lambda_B \cdot t}}{\lambda_B} \right),$$

(1.41)

and so on. The corresponding activities of isotope $A, B, C, \ldots$ can be expressed similarly since $A(t) = \lambda \cdot N(t)$

$$
\begin{aligned}
A_A(t) &= A_A(t = 0) \cdot e^{-\lambda_A \cdot t} \\
A_B(t) &= \frac{\lambda_B}{\lambda_B - \lambda_A} \cdot A_A(t = 0) \cdot \left(e^{-\lambda_A \cdot t} - e^{-\lambda_B \cdot t} \right).
\end{aligned} \tag{1.42}
$$

These equations describe the abundance and/or activity evolution of the various isotopes along the radioactive chain. They state that long-lived isotope—with a small decay constant λ—are enriched, while short-lived isotopes—with a large decay constant λ—only have negligible abundances.

If the decay chain is in equilibrium, this means the depletion of a particular member of the chain equals its production, the decay activity of each member is equal, and the associated abundances N_i correlate with the decay constants λ_i:

$$
\lambda_1 N_1 = \lambda_2 N_2 = \lambda_3 N_3 = \ldots = \lambda_n N_n. \tag{1.43}
$$

This fact may be used to investigate the provenance of certain materials—since the radiogenic composition depends mostly on the geological history of the location where the raw material originated. The intensity of the characteristic γ radiation from decay products along the Uranium or Thorium decay chain in historical pottery or stone artifacts can, for example, provide information on the origin of the material in loess ground or quarries. Since all decay processes follow a characteristic time scale, they offer the unique opportunity of a radioactive clock. The measurement of the characteristic activity $A(t)$ of radioactive material with known half-life $T_{1/2}$ allows us to calculate the age t of the material if the initial amount N_0 is known

$$
t = \frac{T_{1/2}}{\ln 2} \cdot \ln \left(\frac{\ln 2}{T_{1/2}} \cdot \frac{N_0}{A(t)} \right). \tag{1.44}
$$

Since most materials contain a small amount of natural radioactivity, the measurement of its decay provides an ideal tool for dating of ancient artifacts.

1.6 NUCLEAR REACTIONS AND ACTIVATION TECHNIQUES

Rutherford, this is transmutation! For Mike's sake, Soddy, don't call it transmutation. They'll have our heads off as alchemists.

(Ernest Rutherford and Frederik Soddy, exchange 1901)

In the previous section, we discussed the characteristics of nuclear decay processes of unstable radioactive nuclei. While the time scale for some of the decay processes can be in the order of 10^9 years, most of the characteristic life times of radioactive nuclei are considerably shorter. All radioactive isotopes undergo natural decay toward stable daughter isotopes, but new radioactive nuclei can be and are formed through nuclear reactions. Nuclear reactions take place when two (in some cases even three) nuclei interact with each other. Through this interaction process, the particles can change their direction, momentum, and kinetic energy (scattering process), they can transfer energy and change into a so-called excited state (inelastic scattering), or they can change their nature or produce new particles (nuclear reactions) during and through the interaction. Each of these processes has a certain quantum mechanical probability, which depends strongly on the nature of the interaction process. Scattering typically takes place through Coulomb interaction between charged particles, where the particles are deflected through their positive charge. Nuclear reactions that change the proton-neutron configuration in the initial nuclei take place through the strong interaction, and nuclear reactions that exchange the protons and neutrons while maintaining the same mass involve the weak interaction. Nuclear reactions are a very powerful tool with a broad range of applications in modern society. Neutron induced fission reactions generate the energy in nuclear power plants. Nuclear reactions involving radioactive isotope are used for medical applications or, more specifically, for radiation cancer treatment in hospitals. Nuclear reactions find applications in food sterilization and food preservation. Nuclear reactions also enjoy a wide range of analytical applications in material sciences and industry, and last but not least, nuclear reactions are a powerful tool in art analysis as will be discussed in the subsequent sections of this book. While the physics of nuclear reactions is a rather complex subject far beyond the scope of this book, the following section will identify and define some of the underlying principles. These principles are necessary for understanding which nuclear reaction techniques would best be utilized, given their broad range of applications.

1.6.1 REACTION ENERGETICS

In nuclear physics a certain notation is used for describing the large variety of interaction processes. The elastic scattering of projectile on a target nucleus A is described by $A(a, a)A$, since projectile a and target nucleus A only change momentum and kinetic energy but not their configuration. For example, proton scattering on the oxygen isotope ^{16}O would be formulated as ^{16}O$(p, p)^{16}$O. Inelastic scattering is expressed as $A(a, a')A^*$ since the target nucleus is transferred into one of its excited states through energy transfer from the projectile. Inelastic scattering of protons on ^{16}O would be ^{16}O$(p, p')^{16}$O*. The excited state of ^{16}O* is at 6.125 MeV, which would subsequently decay by emission of 6.125 MeV γ radiation. Nuclear reactions are differentiated into capture processes and reaction processes. In the first case, the projectile is captured into the target nucleus, changing the number of protons and/or neutrons in the nucleus; energy is released in the form of γ-emission. Proton capture on ^{16}O is expressed by ^{16}O$(p, \gamma)^{17}$F since the capture of one proton increases the $Z = 8$ (oxygen) to $Z = 9$ (fluorine). In the same way, neutron capture on ^{16}O is expressed by ^{16}O$(n, \gamma)^{17}$O, representing a change along the isotope line since the neutron number is changed but not Z. Some nuclear reactions produce nuclei that cannot be stabilized by mere γ-emission but rather, subsequently break up into smaller nuclei or nucleons. These intermediate and extremely short-lived configurations are called compound nuclei. One example for such a reaction process is proton bombardment of the stable fluorine isotope ^{19}F, which forms the neon compound nucleus ^{20}Ne* in a highly excited state, which immediately decays into an ^{16}O nucleus and a ^{4}He nucleus or α particle. This process is expressed as ^{19}F$(p, \alpha)^{16}$O.

Nuclear reactions that are based on electromagnetic and strong interaction maintain the net number of protons and the net number of neutrons. For the case of the ^{19}F$(p, \alpha)^{16}$O process, the so-called entrance channel ^{19}F$+p$ represents a configuration of $9 + 1 = 10$ protons and 10 neutrons; the compound nucleus ^{20}Ne has 10 protons and 10 neutrons, and the exit channel ^{16}O $+ {}^4$He represents a configuration of $8 + 2 = 10$ protons and $8 + 2 = 10$ neutrons. Weak interaction-based reactions, on the other hand, do not preserve the number of protons and neutrons but only preserve the number of nucleons. This is similar to weak interaction-based decay processes like β-decay or electron capture because these processes convert a neutron into a proton and vice versa. The best-known example for a weak interaction driven reaction is the $p + p \Rightarrow d + e^+ + v$ by which two hydrogen nuclei convert to a deuterium nucleus. This reaction is the first step in the so-called pp-reaction chains, which control energy generation and the lifetime of the sun.

However, not only proton and neutron numbers are conserved; nuclear reactions also conserve mass or energy. The Q-value of a nuclear reaction $A(a, b)B$ represents the mass difference between the total mass of the nuclei in the entrance channel and the nuclei in the exit channel

$$Q = (m_A + m_a) \cdot c^2 - (m_B + m_b) c^2. \tag{1.45}$$

If $Q > 0$ the reaction is exothermic and energy is released in the nuclear process, as kinetic or electromagnetic energy of the reaction products. If $Q < 0$ the reaction is endothermic and requires energy that has to be supplied by the initial kinetic energy of the projectile. The reaction ^{15}N$(p, \alpha)^{12}$C has a Q-value of $Q = 4.966$ MeV, which is calculated from the mass differences between the ^{15}N and ^{1}H nuclei in the entrance channel and the ^{12}C and ^{4}He nuclei in the exit channel. Figure 1.24 shows the energy scheme of the reaction process.

The scheme also indicates that the final nucleus ^{12}C is not produced in its ground state but in an excited state at 4.432 MeV, which subsequently decays by the emission of 4.432 MeV γ-radiation to its ground state. This means that 0.533 MeV of the total released energy will be in the kinetic energy of the α particle and the ^{12}C* nucleus. Another example is the ^{1}H$(p, e^+v)^2$H reaction, where the mass difference between the two protons in the entrance channel and the electron and deuterium in the exit channel yield a Q-value of $Q = 2.224$ MeV—this is a fraction of the energy that is released by the pp-chains in the interior of the sun. Part of that energy is, however, carried away as kinetic energy of the neutrinos. Since neutrinos interact very little with matter they are not absorbed in the solar material and escape unhindered.

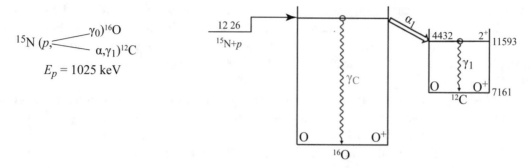

Figure 1.24: Reaction and energy scheme for the 1.025 MeV proton resonance in the two reaction channels $^{15}\text{N}(p,\alpha)^{12}\text{C}$ and $^{15}\text{N}(p,\gamma)^{16}\text{O}$.

1.6.2   REACTION CROSS SECTIONS

As mentioned before, nuclear reactions take place with a certain probability, which can differ for different reactions by many orders of magnitude. The reaction probability is called cross section σ and is traditionally expressed in unit *barns*. The barn reflects the probability that a projectile nucleus can hit a target nucleus of radius r. This probability roughly scales with the areal size of the target nucleus

$$\sigma \approx \cdot \pi r^2 = \pi \cdot r_0^2 \cdot A^{2/3} \approx 5 \cdot A^{2/3} \ [\text{fm}^2]$$
$$\sigma \approx 5 \cdot 10^{-26} \cdot A^{2/3} \ [\text{cm}^2]$$

(1.46)

with $1\ \text{fm} = 10^{-13}$ cm. The mass number ranges from $A = 1$ to $A = 300$, and the unit barn was arbitrarily defined to be

$$1 \text{ barn} \equiv 10^{-24} \text{cm}^2.$$

The probability that two protons ($A = 1$) hit each other is $\sigma = 0.05$ barn, while the probability that two lead isotopes ($A = 208$) hit each other is $\sigma = 1.75$ barn. However, the reaction probability itself does not only depend on the size of the interacting nuclei but also on the quantum mechanical transition probability, which causes the nuclei to change their internal proton-neutron configuration. The cross section also depends sensitively on the energy of the interacting particles, which may change the interaction probability. Interaction between charged particles is largely handicapped by the deflecting Coulomb forces; the cross section drops exponentially with decreasing energy since the particles do not have sufficient energy to overcome the so-called Coulomb barrier. There is no deflecting Coulomb force for interactions with neutrons; in this case the interaction probability increases with decreasing energy. For very slow-moving neutrons the probability for interaction is larger than for fast neutrons, just because of the larger potential interacting time scale between slowly passing nuclei. Figure 1.25 shows the cross section for the $^{16}\text{O}^+\text{p}$ reaction as a function of energy. It can clearly be seen that the cross section drops exponentially toward low energies. In cases of reactions with neutral particles, the Coulomb barrier does not prohibit the reaction process. For neutron capture processes like the $^{16}O^{+}n$ reaction, the cross section therefore increases with decreasing energy or is inversely proportional to the relative velocity of the interacting particles, $\sigma \propto 1/v$; this is known as the $1/v$ law, which applies to most of the neutron induced reaction cross sections. Again, the overall energy dependence of a cross section is determined by the Coulomb repulsion for charged particles and by the $1/v$ law for neutral particles. The absolute magnitude of the cross section depends, however, on the quantum mechanical transition probabilities between the various particle configurations in the entrance and the exit channels. This is a subject that also critically depends on the nuclear structure of the interacting nuclei and the compound nuclei. A more detailed discussion is beyond the scope of this text and we refer to the literature.

Drastic changes in the cross section occur if the reaction energetics are such that a quantum mechanically allowed configuration of the nucleonic compound system is populated. These configurations correspond to energetically excited states of the nucleus. Energetically excited states can be characterized by certain shell configurations of the protons or neutrons in the nucleus that correspond to higher energies than the ground state. Other modes of excitation are the rotational motion of the nucleus, with a certain angular momentum, or a large variety of vibrational modes of excitation. These modes of excitation can populate nuclear reactions if the sum of the reaction Q-value and the center of mass kinetic energy E_p^{cm} of the two interaction particles matches the excitation energy E_x in the compound nucleus

$$E_x = E_p^{cm} + Q = E_p^{cm} + (m_A + m_a) \cdot c^2 - (m_B + m_b) \cdot c^2. \tag{1.47}$$

In such cases, the reaction cross section, for a reaction $A(a, b)B$, can drastically increase with energy and show a strong resonant behavior around the energy E_p^{cm}. The cross section reaches a maximum at the resonance energy E_R. The maximum cross section for such a resonance depends on the quantum mechanical transition probability Γ_{A+a} for the entrance channel of fusion, with the initial two particle configuration $A + a$ reaching the excited compound state configuration, and the subsequent decay probability Γ_{B+b} of this excited state going into the final configuration $B + b$. The resonance cross section is described by the Breit–Wigner with the energy E_p^{cm} in units MeV

$$\sigma(E) = \frac{20.7}{E_p^{cm}} \cdot \omega \cdot \frac{\Gamma_{A+a} \cdot \Gamma_{B+b}}{\left(E_p^{cm} - E_r\right)^2 + (\Gamma_{tot}/2)^2}. \tag{1.48}$$

The total width $\Gamma_{tot} = \Gamma_{A+a} + \Gamma_{B+b}$ corresponds to the total decay probability of the excited state and is inversely proportional to the lifetime of the configuration $\tau = 1/\Gamma_{tot}$. The parameter ω depends on the spin configuration of the resonant state and the interacting particles A and a. The resonant cross section reaches its maximum at the resonance energy

$$\sigma(E_r) = \frac{20.7}{E_p^{cm}} \cdot \omega \frac{\Gamma_{A+a} \cdot \Gamma_{B+b}}{(\Gamma_{tot}/2)^2}. \tag{1.49}$$

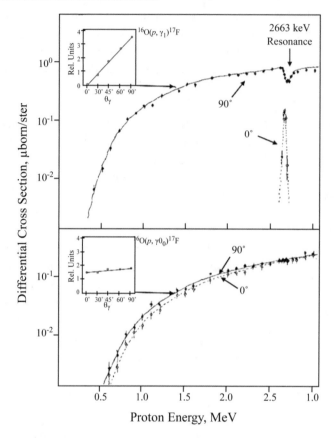

Figure 1.25: Cross section of the $^{16}O(p, \gamma)^{17}F$ reaction as a function of proton bombarding energy. The cross section drops exponentially toward lower energies because of the deflecting Coulomb barrier between the positive charged proton ($Z = 1$) and the positive charged ^{16}O nucleus ($Z = 8$). The inlets show the effects of angular distribution, the dependence of the cross section with detection angle of the emitted γ-radiation.

Figure 1.26 shows a typical resonance cross section for the reaction $^{15}N(p, \alpha)^{12}C$ at a resonance energy of 500 keV. The width of this particular resonance is very broad $\Gamma_{tot} \approx 300$ keV, which represents only a relatively modest change of cross section with energy.

On the other hand, most of the resonances observed in nuclear reactions are very narrow $\Gamma_{tot} \leq 1$ keV, which translates into a very rapid and steep change in cross section. This is, however, difficult to measure since the cross section only resembles a probability function for the nuclear reaction to occur and not a measurable entity. In the case of such narrow resonances, effects of the energy loss of the beam particles in the target material need to be considered. These energy-loss effects will be discussed in more detail in the next section.

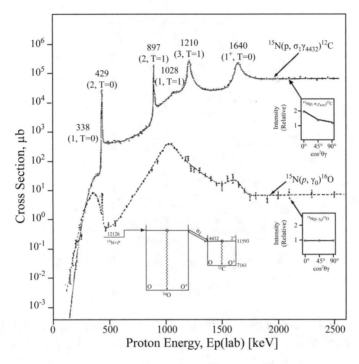

Figure 1.26: Reaction cross sections of the resonant $^{15}N(p, \gamma)^{16}O$ proton capture and the competing $^{15}N(p, \alpha_1)^{12}C^*$ reaction to the first excited state in ^{12}C with subsequent γ-emission to the ^{12}C ground state. The inlets show the angular distribution of the emitted γ-radiation in both processes.

The reaction cross section is the means for labeling the probability that a nuclear reaction will take place between two particles. However, the cross section is not a measurable item; what is measured are the number of reaction products, the so-called reaction yield, in terms of the incoming number of particles, the beam intensity. The yield depends on the particular experimental conditions, while the cross section is independent from the details of the experiment. In a typical nuclear reaction experiment a projectile with certain velocity or kinetic energy is used to bombard a target material.

Figure 1.27 demonstrates the overall probability (cross section) that a single incoming projectile will hit a single nucleus in the target material. If there are many nuclei in the target material, the overall probability Y that the incoming projectile will hit one of them depends on the number N of the target nuclei per volume—the number density—and the thickness of the target material d

$$Y = \sigma \cdot N \cdot d. \tag{1.50}$$

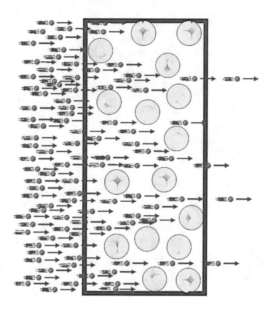

Figure 1.27: The overall reaction probability is called yield Y and depends on the cross section σ but also on the number of particles in the target material N and the thickness d of the targe.

The higher the number of target atoms and the thicker the target, the higher the probability that the incoming particle will undergo some interaction. This probability is called the reaction yield and gives the number of reaction products per incoming particle.

1.6.3   REACTION YIELD

Typically, there is not a single projectile but, rather, a larger number of particles that are penetrating the target material. Taking a projectile beam intensity $I = n_i/t$ of n_i particles per time t, the total number of reactions taking place during a time period t or the total yield N of reaction products per time is

$$N = Y \cdot I = \frac{\sigma \cdot N \cdot n_i \cdot d}{t}. \tag{1.51}$$

This equation determines the number of reaction products originated in the nuclear process. This includes light and heavy particles as well as γ-radiation. In addition, if the nuclear reaction produces a radioactive isotope, the production rate $P = N$ determines how many radioactive isotopes are produced in the activation process.

There is, however, the matter of identifying and counting the newly created nuclear reaction products. This requires particle or photon detectors, which absorb and identify the reaction products. Since a typical detector covers only a small part of all of the directions in which reaction products are emitted, we also have to correct for the so-called solid angle $\Delta\Omega$ of the detector. The solid angle is defined as a function of detector area S and detector distance r from the reaction event; if the detector covers the entire sphere around the reaction event, we have a total solid angle of $\Omega = 4\pi$

$$\Delta\Omega = \frac{S}{r^2}; \quad \Omega = \frac{4\pi \cdot r^2}{r^2} = 4\pi. \tag{1.52}$$

We also need to correct for the solid angle of the detector, which is typically a geometrical problem with regard to the experimental set-up,

$$N_{det}(\Omega) = \eta \cdot Y \cdot I \cdot t = \sigma \cdot n_0 \cdot n_i \cdot d \cdot \frac{\Delta\Omega}{4\pi}. \tag{1.53}$$

Another correction to take into account is the angular distribution of a nuclear reaction process. This helps to identify the optimal position of a detector, in order to maximize the detection of the reaction products.

A nuclear reaction is often characterized by the angular momentum transfer from the interacting particles to the reaction products. The angular momentum coupling between the entrance channel configuration and the exit channel configuration often results in preferred directions for the reaction products—classified by the angle θ with respect to the direction of the incoming particle. This means that the reaction products come with a specific angular distribution $W(\theta)$, which is characteristic for certain angular momenta transfer conditions. To accommodate for this, the differential cross section is defined as the probability that a reaction product is ejected in a direction characterized by the angle θ with respect to the direction of the incoming particle. If the angular distribution is isotropic, reaction products can be detected in all directions. This is important for the measurement and identification of the reaction characteristics. This detection process depends critically on the interaction probability of the particles and photons with the detector material. This interaction probability determines the so-called efficiency η of the detector to absorb and identify a reaction event. This efficiency depends on the interaction cross section and the size or design of the detector. Since the interaction process sometimes has a very small cross section, typical efficiency ranges from $\eta = 10^{-4}$ to $\eta = 1$. The total number of identified reaction products measured for a time t with a specific detector of efficiency η is

$$N_{det}(\theta, \Omega) = \eta \cdot Y \cdot I \cdot t = \sigma \cdot N \cdot n_i \cdot d \cdot \eta \cdot \frac{\Delta\Omega}{4\pi} \cdot W(\theta). \tag{1.54}$$

To use a nuclear reaction efficiently, we need to know the nuclear reaction cross section σ. For optimizing the detection of the reaction products, we need the angular distribution $W(\theta)$ and the detector efficiency η.

1.6.4 PRODUCTION OF RADIOACTIVITY

In the reaction process described above, the initial number of target nuclei are gradually depleted and converted into a different species of isotopes, depending on the reaction mechanism. However, the flux of incoming particles provided by accelerators or reactors is relatively small, at the

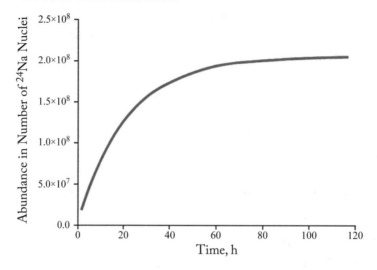

Figure 1.28: The increase of ^{24}Na abundance in the neutron induced activation of a ^{23}Na sample. The equilibrium between production and decay is reached after ten half-lives.

order of 10^{14} particles/s·cm^2, and the cross sections are typically less than 10^{-24} cm^2 (for nuclear reactions). This translates into the probability of converting target nuclei into new species to be less than 10^{-10}/s. Unless one exposes the material to an accelerator beam or reactor flux for an infinite amount of time, the decline in target isotopes is negligible, but one can build appreciable amounts of new isotope material. If the reaction product is radioactive, one produces new activity from the initial number of target nuclei N_0. The production rate R for new radioactive isotopes corresponds to the reaction cross section σ and the beam flux I:

$$R = N_0 \cdot \sigma \cdot I. \tag{1.55}$$

Considering an example of neutron activation of ^{23}Na via the ^{23}Na$(n, \gamma)^{24}$Na reaction, using a sample of 1 μg($N_0 = 2.6 \cdot 10^{16}$ ^{23}Na atoms) with a neutron flux of 10^{14} neutrons/s · cm^2 and a cross section of about 1 mbarn = 10^{-27} cm^2, there is a yield of ^{24}Na with a production rate of about $2.6 \cdot 10^3$/s (Figure 1.28).

Example 1.3   Suppose a ^{12}C target of $d = 0.1$ m thickness is bombarded with a low-energy proton beam of $I = 10\,\mu A$, triggering the reaction ^{12}C$(p, \gamma)^{13}$N with a cross section of $\sigma = 6\,\mu$barn. The reaction product ^{13}N is a β^+ emitter and has a half-life of $T_{1/2} = 9.96$ m. Carbon has a density of 1.8 g/cm^3. According to chemistry, 12 g of ^{12}C contain $N_A = 6.023 \cdot 10^{23}$ ^{12}C atoms (Avogadro's number). This translates into a number density of

$$N = \frac{1.8 \ g/cm^3}{12 \ g} \cdot 6.023 \cdot 10^{23} \ {}^{12}C = 9.0 \cdot 10^{22} \ {}^{12}C/cm^3.$$

An electrical charge of 1 Cb corresponds to a total number of $6.25 \cdot 10^{18}$ single-charged particles since a positively charged proton has the elementary charge of $e^+ = 1.6 \cdot 10^{-19}$ Cb. Therefore, a beam intensity of 10 μA gives

$$\eta_i = 10 \cdot 10^{-6} \ A = 10 \cdot 10^{-6} \ Cb/s = 10 \cdot 10^{-6} \cdot 6.25 \cdot 10^{18} \ proton/s = 6.25 \cdot 10^{13} \ proton/s.$$

This provides a production rate for radioactive ${}^{13}N$ particles and reaction specific γ photons,

$$P = 6 \ \mu b \cdot 9 \cdot 10^{22} \ {}^{12}C/cm^3 \cdot 6.25 \cdot 10^{13} \ p/s \cdot 0.1 \ \mu m$$
$$P = 6 \cdot 10^{-30} cm^2 \cdot 9 \cdot 10^{22} \ {}^{12}C/cm^3 \cdot 6.25 \cdot 10^{13} p/s \cdot 10^{-5} \ cm$$
$$P = 338 \ {}^{13}N \cdot \gamma/s.$$

An activation for 1 h = 3600 s yields a total number of $1.22 \cdot 10^6$ radioactive ${}^{13}N$ particles ($T_{1/2} = 9.96$ m) with an activity of

$$A\left({}^{13}N\right) = \lambda \cdot N\left({}^{13}N\right) \frac{\ln 2}{T_{1/2}} \cdot N\left({}^{13}N\right)$$

$$= \frac{\ln 2}{T_{1/2}} \cdot 1.22 \cdot 10^6 \ {}^{13}N = \frac{\ln 2}{597.6 \ s} \cdot 1.22 \cdot 10^6 decays = 1415 \ Bq.$$

The total β^+-activity of the irradiated sample is 1415 Bq or $3.82 \cdot 10^{-8}$ Ci.

While a radioactive species can be easily produced this way through technical but also natural activation mechanism, the time-dependent abundance is again subject to the decay and needs to be formulated by combining the production and decay equations:

$$\frac{dN_1}{dt} = R - \lambda_1 \cdot N_1$$

$$N_1(t) = \frac{R}{\lambda_1}\left(1 - e^{-\lambda_1 \cdot t}\right). \tag{1.56}$$

With ${}^{24}Na$ having a half-life of $t_{1/2} = 14.96$ h, the ${}^{24}Na$ production reaches an equilibrium in its decay after about eight to ten half-lives. The total amount of the produced activity depends entirely on the production rate.

Example 1.3 demonstrates how to calculate the production rate of ${}^{13}N$ through the radiative capture reaction ${}^{12}C(p, \gamma){}^{13}N$.

1.7   RADIATION MEASUREMENT AND RADIATION EXPOSURE

The actions possessed by the radiations from radio-active bodies of producing charged carriers of ions in the gas, has formed the basis of an accurate quantitative method of examination of the properties of the radiations and of radio-active processes.

(Ernest Rutherford, Radioactivity, 1904)

To study and analyze the radiation emitted in atomic processes, nuclear decay, or nuclear reactions, the radiation must be measured. The type of radiation—X-ray, γ-ray, α-particles, β-particles, etc.—needs to be identified, and its intensity and characteristic energy need to be determined. This requires radiation detectors. Detectors consist of material that has a particularly high interaction probability with a specific type of radiation. The choice of material used for a detector defines its application for γ-, α-, β-, or neutron measurement. Most types of detection material show high sensitivity for more than one kind of radiation. The radiation, which interacts with the detector material, will be efficiently absorbed. The absorbed energy is either released in electromagnetic energy (UV or visible light) or in electrical energy and is subsequently converted to a short electrical pulse. The pulse has a particular shape and a specific height. The shape and duration of the pulse are often characteristic for the radioactive particle that has originated it; the pulse height is directly proportional to the energy of the absorbed radiation photon or particle. The number of pulses is directly proportional to the initial intensity, scaled by the efficiency of the detector material for the particular kind of radiation. There are in principle three kinds of detectors: gas detectors, scintillation detectors, and semi-conductor detectors. These are all based on the basic principles for the interaction of radiation with matter, as outlined in the previous section.

1.7.1   GAS DETECTORS

Gas detectors are particularly sensitive for the detection of charged particles like β- or α-radiation. The particle enters a gas volume and is stopped, ionizing the gas atoms along its trajectory. The released electrons and ions are collected in a strong external electrical field, on two electrodes mounted in the gas, as illustrated in Figure 1.29. This originates an electrical signal at the electrodes, which can either be measured as current in the case of a high-count rate or as an electrical voltage pulse of certain pulse height V.

The pulse height is directly proportional to the number of ion-electron pairs n_0 produced by the ionization event. This number in turn depends strongly on the applied acceleration voltage, or the kinetic energy the particles gain in the accelerating electrical field. The dependence between pulse height and accelerating voltage is shown in Figure 1.30. This illustrates the variation in the number of electron-ion pairs as a function of accelerating potential and also shows the correlation between pulse height and the initial kinetic energy of the radiation particle. The latter defines the conditions for three characteristic applications of ion chambers. The low volt-

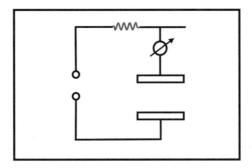

Figure 1.29: Standard scheme for the design and electrical circuitry of a gas-filled ionization chamber.

age region of ion saturation is the range where the electrons have sufficient energy for ionization and the produced electrons are collected without substantial loss through recombination. This is the range used with standard gas-filled ion chamber configurations. At higher voltage, the number of ion-electron pairs increases rapidly since the produced primary electrons can gain sufficient kinetic energy in the accelerating field to initiate secondary ionization processes. In both of these regions, the pulse height or the number of electron-ion pairs is directly proportional to the initial kinetic energy of the radiation particle. In the Geiger–Müller region, the correlation between initial energy of the radiation, or the initial number of pairs produced in the primary ionization event, is lost.

For ionization chambers that operate in the ion saturation range, each ionization event requires on average somewhere between $W_I = 20$ and 35 eV, depending on the kind of gas. Using an average value of $W_I = 30$ eV allows for quickly estimating the number of generated ion-electron pairs per incoming particle with energy E_0.

$$n_0 = \frac{E_0}{W_1} \cong \frac{E_0}{30 \text{ eV}}. \tag{1.57}$$

An α-particle with an initial energy of 3 MeV can therefore ionize up to 10^5 atoms along its trajectory; if the initial α energy is 1 MeV, only $3 \cdot 10^4$ atoms are produced before the particle is fully stopped. The number of ion-electron pairs originated in each stopping event therefore correlates directly with the initial energy of the stopped particle. The migration time of ions to the electrodes is slow and takes a few milliseconds before collection, which prevents single pulse analysis. The electron-ion induced current gives a measure for the intensity N_{rad} of the radiation entering the gas region. Ionization chambers, which operate on the basis of ion-electron pair collection, are therefore frequently used for radiation survey purposes, measuring the radiation induced current I. For the typical charge of $e = 1.6 \cdot 10^{-19}$ C per ion-electron pair, the current

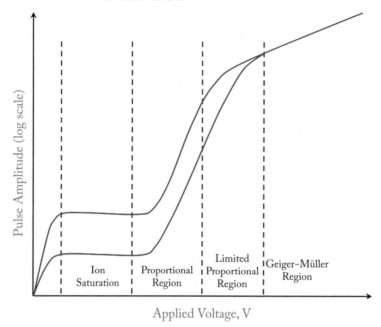

Figure 1.30: Gas response and variation of the output signal from a gas-filled ionization chamber in response to the applied voltage.

is

$$I = N_{rad} \cdot n_0 \cdot e = A(t) \cdot \frac{E_0}{W_1} \cdot e \cong 5.3 \cdot 10^{21} \text{ [C/eV]} \cdot N_{rad} \cdot E_0 \text{ [eV]}. \tag{1.58}$$

An intensity $N_{rad} = 10^4$ s^{-1} of 3 MeV α-particles in the detector translates into an average current of $I = 1.6 \cdot 10^{-10} A = 160$ pA, which is easily measurable.

Electrons have a higher mobility in the gas and therefore can be more efficiently collected at the electrodes, within a few microseconds. If the incoming radiation has a rate of $N_{rad} \leq 10^5$ s^{-1}, the ion chamber can be operated in pulse mode where each incoming radiation particle can be detected separately. In this case not a continuous current is measured but, rather, the voltage pulse originated by the collection of the produced electrons is measured. The associated pulse height is

$$V = \frac{n_0 \cdot e}{C} = \frac{e}{C} \cdot \frac{E_0}{W_I} \cong 5.3 \cdot 10^{-21} \text{ [C/eV]} \cdot \frac{E_0 \text{ [eV]}}{C \text{ [F]}}. \tag{1.59}$$

The electrodes correspond to a capacitor with a capacitance C, which depends on the gas material and the size and shape of the electrodes. A typical capacitance for an ion chamber is $C = 10^{-10}$ F. The pulse height for a 3 MeV α-particle is therefore $V(3 \text{ MeV}) = 1.6 \cdot 10^{-5}$ V $= 16 \ \mu$V and for a 1 MeV α-particle the pulse height is $V(1 \text{ MeV}) = 5.3 \cdot 10^{-6}$ V $= 5.3 \ \mu$V, which are easily distinguishable.

Collection efficiency can be maximized through the geometric design of the electrode configuration. While in principle each kind of gas can be used, the performance can be optimized by the right choice of gas. For example, recombination between the free electrons and positive ions needs to be minimized, otherwise only a small number of the original electrons can be detected and the efficiency of the detector is reduced. Recombination is high in gas combinations that easily form negative oxygen ions, and this favors the use of noble gases like argon, which don't form negative ions at all.

A specific version of a gas-filled detector is the proportional counter, which operates with an accelerating voltage in the proportional range shown in Figure 1.30. With the proportional counter, a special amplification effect is achieved by significantly increasing the accelerating electrical field for the electrons until they gain sufficient kinetic energy for secondary ionization. Primary and secondary electrons produce further ionization, which induces a rapid amplification by causing an electron avalanche of total charge Q with typical multiplication factors of $M = 10^3$ to 10^5 for each initial electron:

$$Q = n_0 \cdot e \cdot M. \tag{1.60}$$

This increases the pulse height $V = Q/C$ at the collecting electrode (anode), but within a certain electrical field range the pulse height remains proportional to the number of primary electrons n_0 and therefore proportional to the energy of the initial radiation event. To achieve the necessary high field strengths, proportional counters are usually designed in a cylindrical configuration where the electron-collecting anode consists of a wire of radius a stretched along the cylinder axis and the ion-collecting cathode is a concentric cylinder of radius b around the wire. The resulting radial electrical field $E(r)$ depends on the geometrical dimensions of the system and on the applied voltage V between the cylindrical cathode and anode wire

$$E(r) = \frac{V}{r \cdot \ln(b/a)}. \tag{1.61}$$

With typical dimensions of $a = 0.01$ mm, $b = 20$ mm, and a voltage of 2 kV, a field strength of $E(r) = 263/r$ [V·m^{-1}] is achieved. The multiplication takes place in the region of maximum field strength, close to the central wire, which for this example is estimated to $E(a) \approx 2.6 \cdot 10^7$ V·m^{-1}. The overall multiplication factor itself can also be approximated

$$\ln M = \frac{V}{\ln(b/a)} \cdot \frac{\ln 2}{\Delta V} \cdot \left[\ln \frac{V}{p \cdot a \cdot \ln(b/a)} - \ln K \right] \tag{1.62}$$

and depends on the geometry of the proportional counter, the pressure p in the counter gas (in units atmosphere) and the quantities K and ΔV, which are characteristic for the specific gas mixture used. Typically, values range between $K \approx 5 - 7 \cdot 10^{-4}$ [V/cm-atm] and $\Delta V \approx 22 - 28$ [eV] for an Argon-methane gas mixture.

The Geiger–Müller counter is typically based on a cylindrical design in order to achieve the high accelerating potentials necessary to operate in its specific mode. The energy of the accelerating particles is high enough to excite electrons from the inner shells of the gas atoms.

De-excitation causes the emission of ultra-violet light to X-rays, which in turn cause further ionization and excitation, creating an avalanche of charged particles. This is illustrated schematically in Figure 1.30. Due to the complexity of the secondary ionization processes, the memory of the initial number of electron-ion pairs produced by the incoming radiation is lost and the signal is directly proportional to the applied voltage independent of the energy and nature of the incoming radiation. This is the application range for so-called Geiger–Müller counters, which mainly serve as radiation monitors and survey instruments.

1.7.2   SCINTILLATION DETECTORS

Scintillation detectors are based on the conversion of the incoming radiation and radiation energy into light. This process takes place through the energy loss of the incoming radiation in multiple scattering or other attenuation events. The initial energy is dissipated into excitation of atomic or molecular electron configurations, which subsequently de-excite by emission of light in the visible or near visible UV to IR range. The particular frequency range depends on the scintillation material. Modern scintillators are transparent to characteristic fluorescence light in order to reduce internal absorption, which would reduce the overall efficiency of the detection process. This is achieved by guiding the fluorescence light and directing it by reflection onto a photosensitive surface, the photo cathode. By photo effect, photoelectrons are produced and are accelerated by an externally applied voltage in the so-called photomultiplier. The photomultiplier consists of a series of electrodes, the dynodes, which are connected by a resistor chain to establish an accelerating potential gradient between the dynodes. The electrons are accelerated, by these potential differences, toward the electrodes and they release a multitude of secondary electrons when hitting the dynode. This causes a multiplication effect from the initial number of photoelectrons to the number of electrons released at the anode, the last electrode in the dynode chain. The principle of the scintillation process and the photomultiplier is shown in Figure 1.31. Each dynode represents an average amplification of $A_i = 5$ in number of electrons. In a photomultiplier tube consisting of 10 dynodes, this translates into an overall amplification of $A_{tot} = 5^{10} \approx 10^7$ in number of electrons. The overall efficiency of the scintillation detector depends on the light output from the scintillation material and the efficient multiplication of the photoelectrons to generate a suitable electrical signal at the anode of the phototube. The sensitivity of the detector for different kinds of radiation depends on the specific scintillation material, the nature of the interaction, and the energy loss effects of the radiation.

There are in general two kinds of scintillation material, organic, and inorganic. The first is based on the excitation of molecular electronic or vibrational states in a complex organic material. The excitation energy ranges between 0.1 and 10 eV; this means that absorption of a 1 MeV particle generates 10^6–10^7 excited molecular configurations. According to Equation (1.1), these configurations de-excite by the emission of $E = 0.01$ eV–10 eV photons in the wavelength range

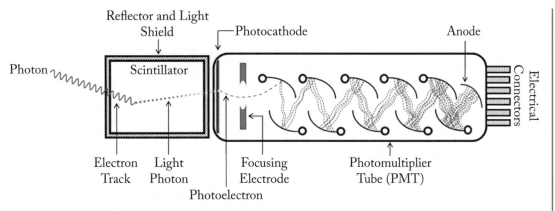

Figure 1.31: Scintillator assembly with scintillator material, photo-cathode, and photomultiplier tube for originating a strong electrical signal.

of

$$\lambda = \frac{c}{v} = \frac{h \cdot c}{E} = \frac{4.136 \cdot 10^{-15} \, [\text{eV} \cdot \text{s}] \cdot 2.998 \cdot 10^{10} \, [\text{cm/s}]}{E \, [\text{eV}]} = \frac{1.24 \cdot 10^{-4}}{E \, [\text{eV}]} \, \text{cm.} \qquad (1.63)$$

This translates, for our example, to a wavelength range of roughly $\lambda \approx 100$ nm to 10 μm, covering the visible to infrared range as shown in Figure 1.1.

The light output dL/dx for organic scintillation materials depends on the energy loss of the radiation dE/dx and the scintillation efficiency $dL/dE = \eta_L$, which represents the excitation probability in the molecular configurations and is described in Birk's formula

$$\frac{dL}{dx} = \frac{\eta_L \cdot \frac{dE}{dx}}{1 + kB \cdot \frac{dE}{dx}}. \qquad (1.64)$$

The parameter kB is a material constant for the scintillator, expressing the ionization density along the radiation track, which reduces the probability of forming molecular configurations. For high-energy β radiation or γ-radiation, the change of energy loss dE/dx is small. This suggests that the total light output from the scintillator is directly proportional to the energy of the radiation

$$L = \int_0^E \frac{\eta_L \cdot \frac{dE}{dx}}{1 + kB \cdot \frac{dE}{dx}} .dx \approx \int_0^E \frac{dL}{dE} \cdot dE = \eta_L \cdot \int_0^E dE = \eta_L \cdot E. \qquad (1.65)$$

For massive particles like α-particles, the energy loss is large, causing a high ionization probability along the radiation track; further, the energy loss is large particularly toward lower energies

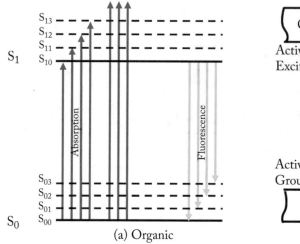
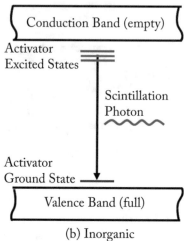

(a) Organic   (b) Inorganic

Figure 1.32: Typical band structure for (a) an organic scintillator with many possible molecular levels and closely spaced vibrational states and (b) an inorganic scintillator with activator states within the energy gap of the pure crystal.

where the α-particles are stopped at the range R

$$L = \int_0^E \frac{\eta_L \cdot \dfrac{dE}{dx}}{1 + kB \cdot \dfrac{dE}{dx}} \cdot dx \approx \int_0^E \frac{\eta_L \cdot \dfrac{dE}{dx}}{kB \cdot \dfrac{dE}{dx}} \cdot dx = \frac{\eta_L}{kB} \cdot R. \tag{1.66}$$

The total light output L is directly proportional to the range R of the particle, which in turn correlates directly with its initial energy, as seen with Equation (1.65).

Inorganic scintillation material does not show the relatively complex molecular structure that characterizes the organic material and leads to the frequent formation of molecular modes of excitation. Inorganic scintillators are typically crystal lattices like NaI or CsI with metallic impurities embedded in the lattice. In a pure lattice, the atomic energy levels overlap and form the so-called band structures of the crystal, representing allowed electron configurations as illustrated in Figure 1.32.

The valence band is the highest energy range for atomic electrons, from the ions forming the crystal lattice. The next band is the conduction band, which is typically void of electrons in the case of an insulating material; electrons in this band would comprise the material and electrical conductor. The two bands are separated by an energy gap, which represents configurations that electrons are quantum mechanically not allowed to populate. If the lattice has impurities, like for example Thallium (Tl) atoms, the band structure is locally deformed and allowed configurations are introduced, which are energetically positioned between the valence and the conduction band.

These energy levels are the activator states. If radiation interacts with the lattice material, energy is absorbed and electrons are excited from the valence band configuration to the conduction band configuration. This leaves an empty configuration in the valence band, which is called *hole* configuration. The hole will be filled with an electron from an activator state, since that resembles an energetically more favorable (lower) configuration. De-excitation of the excited electron in the conductor band follows, through cascade decay from the conduction band to the hole in the activator state, with emission of photons. Given the right choice of activator material, the energy difference is such that the energy of the photons corresponds to the energy range for visible light in the wavelength range of $\lambda \approx 390$–570 nm.

The typical light output for inorganic scintillation material depends on the probability of forming electron-hole pairs and de-excitation through the activator states. The typical band gap width in a scintillation material is $\Delta E_{gap} \approx 20$ eV. The creation of an electron-hole pair therefore requires an average energy of $E \geq 20$ eV. For a radiation particle with an energy of $E_0 = 1$ MeV, about $5 \cdot 10^4$ electron-hole pairs are created, which de-excite by ≈ 3 eV photon emission through the activator states in the optical range around

$$\lambda = \frac{c}{v} = \frac{h \cdot c}{E} = \frac{4.136 \cdot 10^{-15} \text{ [eV} \cdot \text{s]} \cdot 2.998 \cdot 10^{10} \text{ [cm/s]}}{E \text{ [eV]}} = \frac{1.24 \cdot 10^{-4}}{E \text{ [eV]}} \text{ [cm]} \approx 400 \text{ [nm]}.$$

(1.67)

The advantage compared to organic scintillators is that the photon range is much narrower and, due to the choice of activator material, is specifically designed to be in the particular wavelength range of visible light. The broad wavelength range associated with the use of organic scintillators reduces the overall number of photons in the visible wavelength range to a fraction of the number of photons produced in inorganic scintillation materials. The properties of typical inorganic scintillator materials re listed in Table 1.5

The light output dL/dx for inorganic scintillation detectors is less energy dependent than in the case of organic material. In NaI and CsI material, the light output is essentially constant for γ radiation above 400 keV. Deviations only occur at lower energies, due to the excitation of inner shell electrons in the iodine. This is illustrated in Figure 1.33, a comparison of the light output in a NaI scintillation detector and in a NE-102 organic scintillator, for totally absorbed electrons as a function of electron energy.

An important component of the scintillation detector is the photomultiplier, which is directly, optically coupled to the scintillation material. The light is collected from the scintillator with a certain efficiency, which mainly depends on the geometry of the detector and the internal reflectivity of the material. The collected light is converted at the photocathode to electrons, by the photoelectric effect (see Equation (1.67)). The number of released photoelectrons compared to the number of incident photons is expressed in terms of the quantum efficiency $\eta(\lambda)+$. The quantum efficiency depends on the photocathode material. As illustrated in Figure 1.34, the quantum efficiency for various photocathode materials peaks to $\eta(\lambda) \approx 0.2$ in the optical range

Table 1.5: Properties of the inorganic scintillator materials

Material	Wavelength of Maximum Emission (nm), λ_m	Decay Constant (μs)	Index of Refraction at λ_m	Specific Gravity
NaI(Tl)	410	0.23	1.85	3.67
CsI(Na)	420	0.63	1.84	4.51
CsI(Tl)	565	1.0	1.80	4.51
^{6}LiI(Eu)	470–485	1.4	1.96	4.08
ZnS(Ag)[a]	450	0.20	2.36	4.09
CaF$_2$(Eu)	435	0.9	1.44	3.19
Bi$_4$Ge$_3$O$_{12}$	480	0.3	2.15	7.13
CsF	390	0.005	1.48	4.11
Li glass[c]	395	0.075	1.55	2.5

for $\lambda \approx 400$ nm, and the sensitivity of the photomultiplier should be therefore optimized to that wavelength range.

With an incident number of 10^5 photons, the number of released electrons is $2 \cdot 10^4$. To generate an appreciable electrical signal, the number of electrons must be substantially increased by a gain factor of approximately 10^6. This is achieved by electron multiplication in the photomultiplier. The photomultiplier consists of N dynodes (typically $N = 10$–15) with an accelerating potential difference V_d between the dynodes. The electrons are accelerated toward the dynode and excite new additional electrons. The typical gain factor of electrons depends on the electron energy or the accelerating potential $\delta = \eta_d \cdot V_d$ (with the proportional factor η_d); a typical value is $\delta \approx 5$ per stage. If the total voltage V applied to the photomultiplier tube is equally divided between the N dynode sections $V = N \cdot V_d$, the overall gain G is

$$G = \delta^N = (\eta_d \cdot V_d)^N . \tag{1.68}$$

For a typical photomultiplier with $N = 10$ dynode stages and a gain factor of $\delta \approx 5$, the total gain is $G \approx 10^7$. With a typical total voltage between $V = 800$ and 1500 V the potential difference between the various dynodes is $V_d \approx 50$–100 V, depending on the number of dynodes. The last electrode in the photomultiplier device is the anode. The output signal at the anode is again either a current or a charge pulse, with an amplitude directly proportional to the initial number of electrons emitted by the photo cathode and to the multiplication factor G of the photomultiplier itself.

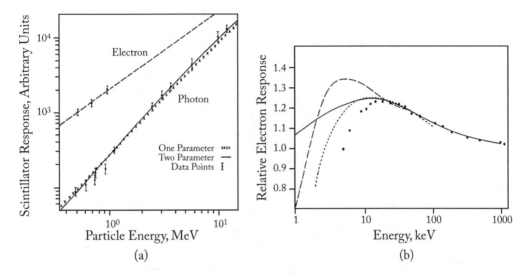

Figure 1.33: Scintillator light output for a plastic scintillator material NE-102 (a) and for an organic NaI crystal (b), as a function of particle/photon energy.

1.7.3 SEMICONDUCTOR DETECTORS

The gas detector and the scintillation detector are based on signals associated with the generation of an electron avalanche brought about by conversion of the initial particle energy through ionization processes and the origination of the electron flux through the photo effect. While these processes generate a sufficiently large electrical signal, which can be detected as the characteristic signature for a nuclear event, the associated statistics, of an initially small number of electrons or photons N, limit reliability in the energy determination of the original event. The energy can only be determined with limited resolution

$$R = \frac{\Delta E}{E} \approx \frac{2.35}{\sqrt{N}}. \tag{1.69}$$

For a typical ionization detector as discussed in Section 1.7.1, the number of electron-ion pairs produced by an α particle of $E = 1$ MeV energy is $N = 3 \cdot 10^4$; this translates into a resolution of $R \approx 1.3\%$ or a peak width of $\Delta E \approx 13$ keV. Similar values for resolution are obtained with scintillation detectors. The semiconductor detector typically has a resolution of $R \approx 0.2$–0.6% depending on radiation type, which is better by nearly an order of magnitude when compared with the resolution of the previously discussed detector types. There are some disadvantages, as far as dimension and costs are concerned, but semiconductor detectors have emerged within the last two decades as the major detector tool for a broad range of art analysis techniques.

The basic principle of semiconductors is similar to that of an ionization gas counter except that the stopping medium for the radiation is a semiconductor solid material. Typical semicon-

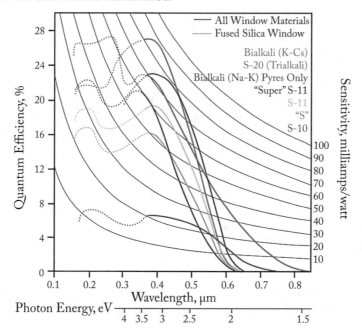

Figure 1.34: The spectral sensitivity or quantum efficiency for a number of photo cathode materials used in photomultiplier tubes.

ductors are Germanium (Ge) and Silicon (Si), in a pure state crystalline material with only a very small energy gap of $\Delta E_{gap} \approx 1$ eV between the valence band and the conductor band. Radiation entering the crystal is being attenuated or stopped and creates a large number of electron-hole pairs along its trajectory. Because of the small band gap, a much larger number of electron-hole pairs are produced than the number of electron-ion pairs in gas detectors, or the number of electron-hole pairs in inorganic semiconductors where each pair creation requires an energy transfer of ≈ 30 eV or 20 eV, respectively. For a 1 MeV α particle the number of pairs $N = 10^6$, according to Equation (1.69), gives a resolution of $R \approx 0.23\%$. An electrical potential ($V \approx 2000$ V) is applied to the crystal, which separates the electrons and holes (which behave like positively charged particles). The electrons are collected at the anode of the detector in a manner similar to the process discussed in the case of gas ionization chambers.

Impurities in the semiconductor material might reduce the band gap width creating a fair amount of thermally excited electron-hole pairs, which induce a steady current when an external electrical field is applied. This current prevents the detection of any radiation event induced electron-hole pairs. Therefore, a special semiconductor design is required to reduce the formation of a charge, with a carrier-free volume thick enough to stop the radiation particles.

Table 1.6: Exposure rate constants for standard radioactive sources

Source	ϕ [R cm²/h mCi]
^{137}Cs	3.3
^{57}Co	13.2
^{22}Na	12
^{60}Co	13.2
^{222}Ra	8.25

Two specific types of semiconductor detectors have been developed, the p–n junction, which is mainly used for particle detection, and the intrinsic detector for photon detection.

1.7.4   BASICS OF DOSIMETRY

Dosimetry is concerned with the quantity of radiation received by an object. It is therefore an important aspect for the safety considerations of radiation exposed workers, for nuclear medicine applications, but also for specific applications in the field of material analysis and material dating in archaeometry. The quantity of radiation received by an object is either expressed in terms of the quantity of subsequent ionization or in terms of the amount of energy the radiation has transferred into the material. The classical unit for measuring the amount of ionization is the *Roentgen*; 1 R (Roentgen) is defined as the amount of exposure that will create a charge of $2.58 \cdot 10^{-4}$ C of singly charged ions in 1 kg of air. The average energy for creating a single ion ($1.6 \cdot 10^{-19}$ C) in air is $W_e \approx 34$ eV. Since the exposure to 1 R of radiation creates $1.6 \cdot 10^{15}$ ions in 1 kg of air, the corresponding total energy absorption by air is therefore $5.5 \cdot 10^{16}$ eV/kg or $8.8 \cdot 10^{-3}$ J/kg. To estimate the effect of radioactive sources with specific activity A, the exposure rate can be expressed in terms of activity A and distance d between the radioactive source and the exposed object

$$\frac{dR}{dt} = \frac{\phi \cdot A}{d^2}. \tag{1.70}$$

The exposure constant ϕ depends on the energy of the radiation, which defines the ionization probability. Table 1.6 lists the exposure constants for several well-known radioactive sources in units $[R \cdot cm^2/h \cdot mCi]$.

Because of the difficulty in applying the *Roentgen*—for various materials with different ionization effects, relative to the various kinds of radiation particles—a new standard has been introduced, which measures the exposure in terms of the amount of energy absorbed per unit mass of material. This quantity is called the *absorbed dose* (D) with a unit of 1 rad = 100 erg/g = 10^{-5} J/g = 10^{-2} J/kg. With the introduction of the international system for standard units (SI), the unit *rad* was officially replaced by the unit *Gray* (Gy), 1 Gy = 1 J/kg = 100 rad, but *rad* can still be found frequently in the literature. An exposure of 1 R gives an absorbed

Table 1.7: Quality factor for the different kinds of particle and gamma radiation

Radiation Type	γ	β	α	Protons	Slow n	Fast n
Q	1	1	20	10	3	10

dose of $D = 0.88$ rad $= 0.0088$ Gy in air. Due to the difference in ionization potential between various materials, the conversion factor is material dependent. Since the average ionization energy for body tissue is $W_I \approx 36$ J/C, an exposure of 1 R corresponds to a dose of $D \approx 0.96$ rad $= 0.0096$ Gy.

Exposure, exposure rate, and absorbed dose are independent of the nature of radiation. However, the damage caused by radiation on body tissue depends strongly on the kind of radiation the body is exposed to. This is mainly due to the energy transfer mechanism or the energy loss dE/dx of the incoming particle in the body material. Since the energy loss of charged particles such as α radiation is considerably higher than for β- or γ-radiation, the effective damage for 3 MeV α-particles is 20 times as high as the damage done by 3 MeV β-particles or 3 MeV γ-photons. To better assess the biological effects for the different kinds of radiation, a new empirical unit has been introduced, the *dose equivalent* (H), which assesses the radiation impact by a quality factor Q

$$H = Q \cdot D. \tag{1.71}$$

The quality factor depends on the radiation of the energy; typical quality factors for radiation with energy below 5 MeV are given in Table 1.7. The unit for the dose equivalent is the *Sievert* (Sv), which replaces the classical unit rem.

$$1 \text{ [Sv]} = Q \cdot 1 \text{ [Gy]}; \quad 1 \text{ [rem]} = Q \cdot 1 \text{ [rad]}; \quad 1 \text{ [Sv]} = 100 \text{ [rem]}.$$

Using the dose equivalent normalizes the effects of the different kinds of radiation to each other; for example, 1 rem of α particles has the same biological effect as 1 rem of γ-radiation. The dose equivalent for an absorbed γ-dose $D = 0.575$ Gy is $H = 1 \cdot 0.0575$ Sv. Since the absorbed dose does not depend on the nature of the radiation, but only on the activity of the source and the time of exposure, the absorbed dose in both cases is $D = 0.575$ Gy. The dose equivalent for the α exposure is $H = 20 \cdot 0.0575$ Sv $= 1.15$ Sv and the dose equivalent for the exposure to thermal neutrons is $H = 30.0575$ Sv $= 0.17$ Sv. Example 1.4 demonstrates how to estimate the dose one is exposed to working with radioactive sources.

Example 1.4   What is the absorbed dose you receive by working for 2 hours at an average distance of 2 m from a ^{22}Na source with an activity of $A = 100$ μCi.

$$\frac{dE}{dt} = \frac{12 \cdot 0.1}{200^2} = 3 \cdot 10^{-5} \text{ [R/h]}.$$

After 2 hours, the exposure is 0.06 mR. This corresponds to an absorbed dose of $D = 0.06 \, [R] \cdot 0.0096 = 5.75 \cdot 10^{-4} \, [Gy] = 0.0575 \, [rad]$.

Example 1.5   Determine the dose equivalent for the ^{22}Na source of Example 1.4; compare it with the equivalent dose from an alpha source and a source of thermal neutrons of the same activity $A = 100 \, \mu$Ci after 2 hours of exposure.

1.8   NATURAL RADIOACTIVITY AND RADIATION EXPOSURE

Results of these observations appear to be best explained by the assumption that a radiation of very high penetrating power enters our atmosphere from above...

(Victor F. Hess, 1912)

Radioactivity is a natural phenomenon. We are exposed to radioactive isotopes all around us, in our daily environment and within us, distributed in our body material. We breathe and eat radioactive substances every day; we exhale these and find them in our excrement. The average amount of radioactivity in our bodies is based on the equilibrium of these processes and on our environmental exposure. The total annual dose equivalent from natural radiation exposure is $H_{tot} \approx 2600 \, \mu$Sv or 260 mrem. The majority of this radioactivity ($\approx 85\%$) is due to natural production processes beyond our control, however, an increasing fraction of our exposure to radioactive isotopes is due to modern technological developments and their impact on our daily lives. In the following, different sources of natural radioactivity are discussed; these are the backbone for the majority of dating techniques of archaeological and anthropological artifacts. The increase in man-made radioactivity, ranging from nuclear bomb tests to medical diagnostics, has an impact on the application of these analysis techniques and is in fact increasingly utilized in modern applications. We therefore will also present a short overview of man-made radioactive exposure.

1.8.1   COSMOGENIC RADIATION

There are two kinds of natural radioactivity, cosmogenic radioisotopes, and radiogenic radioisotopes. Cosmogenic nuclei are produced by interaction between the incoming flux of cosmic ray particles and the atoms and nuclei in the atmosphere as well as in the upper layers of the earth's crust. Cosmic rays are high-energy particles of extraterrestrial origin. They are called primary cosmic rays, originating from our sun and from distant sources far outside our solar system. Primary cosmic rays interact with nuclei in the outer atmosphere of the earth and create, through nuclear reaction and spallation processes, new and often long-lived radioactive particles

Figure 1.35: Cosmic ray air shower. The initial high-energy cosmic ray particles are produced either by solar eruptions or explosions of far distant stars. Interaction with particles in the atmosphere causes spallation and fragmentation processes leading to the development of a multi-particle air shower, with the initial energy being distributed over the entire particle assembly.

like ^{3}H ($T_{1/2} = 12.3$ y), ^{7}Be ($T_{1/2} = 53.3$ d), ^{10}Be ($T_{1/2} = 1.51 \cdot 10^6$ y), ^{14}C ($T_{1/2} = 5730$ y), ^{26}Al ($T_{1/2} = 7.2 \cdot 10^5$), ^{39}Ar ($T_{1/2} = 269$ y), and ^{129}I ($T_{1/2} = 1.57 \cdot 10^7$ y) to name only a few. Through these interactions, the initial energy of the primary cosmic ray particle is converted into the formation of new particles, forming a cascade or shower of muons, leptons, and hadrons. These so-called secondary cosmic rays may also contribute to the production of long-lived cosmogenic nuclides, through spallation processes in the lower atmosphere, and "in situ" in the upper rock layers at the earth's surface. Figure 1.35 illustrates the evolution of a so-called cosmic ray cascade, triggered by the entrance of a high energy primary cosmic ray proton into the outer layer of the earth's atmosphere.

Primary cosmic rays contain a high flux of protons, α-particles, and a small fraction of higher mass nuclei, ranging from beryllium, boron, and carbon, up to iron isotopes. There is

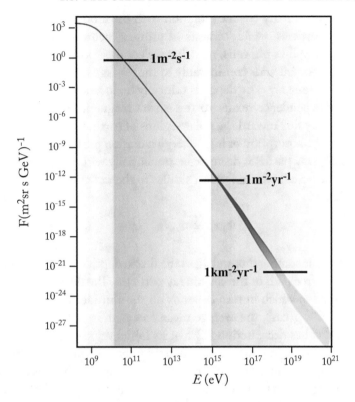

Figure 1.36: The flux of cosmic ray particles as a function of their energy. The flux for the lowest energies (yellow zone) is mainly attributed to solar cosmic rays, intermediate energies (blue) to galactic cosmic rays, and the highest energies (purple) to extragalactic cosmic rays.

also a small flux of γ-photons, electrons, and positrons. Primary cosmic rays come with a broad energy distribution, as shown in Figure 1.36. The distribution has a maximum flux of relatively low-energy particles up to a few MeV and falls off exponentially with increasing energy, up to 10^6 GeV energies. Beyond this so-called *knee* of the distribution, we observe the very high energy range of cosmic ray particles, characterized by a rapid decline in flux up to 10^{10} GeV.

Cosmic radiation is classified according to its origin. Anomalous cosmic rays consist of a flux of relatively low-energy particles (< 10 MeV), which are accelerated in the magnetic field of the sun. Solar cosmic rays in the energy range of 10–100 MeV originate from solar flares and the flux depends on internal solar activity. Galactic cosmic rays are a consequence of supernova explosions within our galaxy and are accelerated through interaction with the magnetic field of the expanding supernova shock front. The origin of the very high-energy cosmic rays beyond 10^{10} GeV is not yet known. These are the highest energy particles observed in nature and have only been discovered within the last two decades.

The interaction of primary cosmic radiation with the atoms in our atmosphere creates a cascade of secondary particles, which consists of protons, neutrons, pions, and other strong-interacting particles, as well as photons, neutrinos, electrons, positrons, and muons. This secondary cosmic ray component adds considerably to the average exposure rate of human beings. However, the annual cosmic ray dose depends critically on location, specifically on latitude and altitude. The latitude dependence relates to the earth's magnetic field, which funnels a larger fraction of the cosmic ray flux toward the polar regions of earth. The increase of cosmic ray flux with altitude is due to the absorption of high energy radiation present in the earth's atmosphere. The total flux of cosmic ray particles decreases exponentially with penetration depth x into the atmosphere and can be described in a fashion similar to the radiation absorption laws discussed in the previous sections

$$\Phi(x) = \Phi_0 \cdot e^{-x/\Lambda} \tag{1.72}$$

with Φ_0 being the initial incoming flux and $\Phi(x)$ the flux at depth x (g/cm2) while $\Lambda(160 \ \text{g/cm}^2)$ is the interaction mean free path of the cosmic ray particles. The decline in flux depends on the density in the atmosphere, which in turn depends on the altitude. With an average density for air of approximately 10^{-3} g/cm^3, the average mean free path is ≈ 1.6 km and at a penetration depth of 10 km, the flux has decreased to 0.2% of its initial value.

At sea level the cosmic ray flux is dominated by muons with only a fraction of γ-rays, neutrons and electrons. The more massive protons and α-particles have been absorbed in high altitude, due to their large energy-loss effects. Figure 1.37 illustrates the equivalent dose rate for different cosmic ray particles as a function of altitude.

With an equivalent dose rate of 0.032 μSv/h, the annual equivalent dose rate from cosmic rays at sea level is $H_{cr} \approx 280 \ \mu$Sv; this is roughly 10% of the total exposure to natural radiation sources. Radiation exposure roughly doubles with every 1500 m in altitude. For high-altitude mountain (≈ 3 km) inhabitants, the annual equivalent dose from cosmic radiation is $H_{cr} \approx 1120 \ \mu$Sv, which increases the annual total exposure considerably. This also has an impact on the radiation exposure of transoceanic flight passengers and aircrews. With a typical flight altitude of 12 km, the radiation dose rate is approximately 0.51 μSv/h, which is more than 10 times the dose rate at sea level. Frequent high-altitude flights may therefore increase the annual dose rate significantly (Example 1.6).

Example 1.6 Calculate the annual dose of a frequent flyer passenger with an average of 5 transatlantic round trips per year at a flight height of 15 km.

The time per transatlantic flight is ≈ 10 h. According to Figure 1.37, the equivalent dose rate is 10 μSv/h, the total equivalent dose for one round trip is $H = 200 \ \mu$Sv, subsequently the total equivalent dose for five round trips is 1000 μSv. Compared to the

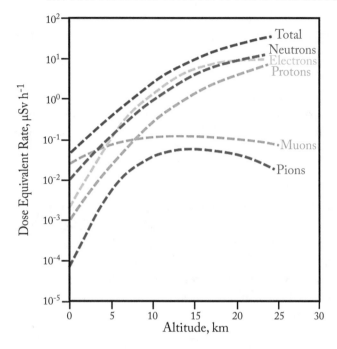

Figure 1.37: The equivalent dose rate for different cosmic ray particles as a function of altitude in the earth's atmosphere.

annual exposure at sea level of $H_{tot} \approx 2600\ \mu$Sv, the passenger increases the annual dose of radiation by one third.

Of particular concern is the production of long-lived radioactive nuclei by cosmic ray interaction processes. These cosmogenic nuclei are created either through spallation processes of very high-energy primary particles or through secondary proton, neutron, or muon induced reaction processes. On average, high-energy neutrons participate in two to three spallation reactions before undergoing nuclear reactions. High-energy protons lose more energy through Coulomb interaction and ionization, and participate on average in only one spallation process. Therefore, most of the production of cosmogenic isotopes in the atmosphere takes place through neutron induced processes at an altitude of about 15–20 km, where the neutron flux reaches its maximum. At ground level, most radioactive isotope production takes place through muon induced interactions, since muons are the dominant remaining cosmic ray component at sea level.

The total production rate $Px(h, N)$ for a particular radioactive isotope x is a rather complex function that depends on a large number of different interaction processes between the cosmic ray particles and the target elements in the earth atmosphere. As pointed out earlier, the production rate P of a certain isotope by a single reaction depends on the flux of the cosmic projectiles Φ, the number of target nuclei N, and the energy-dependent cross-section $\sigma(E)$ for the reaction

$$P = \Phi \cdot \sigma(E) \cdot N. \tag{1.73}$$

Since the atmosphere contains many isotopes, different reactions between the cosmic ray particles and the various isotopes N_j can take place and contribute to the production rate with cross sections σ_j. The total production rate is the sum of all of the contributing processes

$$P = \sum_j \Phi \cdot \sigma_j(E) \cdot N_j. \tag{1.74}$$

Strictly speaking, the abundances of target isotopes N_J in the atmosphere vary with time, particularly with regard to paleoclimate conditions, which are correlated with substantial differences in atmospheric composition when compared to what is observed today.

The cosmic ray flux contains a multitude of high-energy particles with flux Φ_k which can all contribute to the production rate by different interaction processes σ_{jk}. The total production rate therefore requires taking into account all possible cosmic ray flux components, typically neutrons, protons, and muons and is expressed by

$$P = \sum_k \sum_j \Phi_k \cdot \sigma_{jk}(E) \cdot N_j. \tag{1.75}$$

Cosmic rays are not mono-energetic, as shown in Figure 1.36 and the associated reaction cross sections can vary considerably with energy. To determine the total flux we therefore have to integrate over the entire energy distribution of the cosmic ray flux $\Phi_k(E)$. We also have to take into account that the density and therefore the number of target nuclei $N_j(h)$ varies with altitude h. As pointed out before, due to the absorption processes the flux for the different cosmic ray components depends critically on the altitude and varies also with latitude λ, due to the influence of the earth's magnetic field, $\Phi_k(E, h, \lambda)$.

As shown in Figure 1.38, the cosmic ray flux does also vary with time t, due to the influence of the time-dependent behavior of the solar and earth's magnetic fields, which has to be taken into account. This includes variation in the level of solar activity over larger periods of time, which can be due to magneto-hydrodynamic processes in the solar atmosphere but also to the characteristics of the earth-solar field. In addition, there are the 11-year flux modulations that are associated with solar sunspot activity. All of these processes cause variations in the ^{14}C production rates and need to be taken into consideration.

The total production rate for a cosmogenic isotope x is therefore expressed as a function of altitude, latitude, and time in the following formula:

$$P_x(h, \lambda, t) = \int_E \sum_k \sum_j \Phi_k(E, h, \lambda, t) \cdot \sigma_{jk}(E) \cdot N_j \cdot dE. \tag{1.76}$$

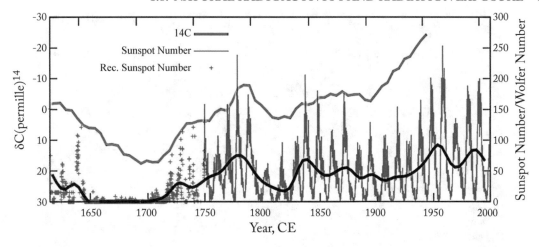

Figure 1.38: The correlation between solar activity as reflected in the average number of sunspots, and variations in the ^{14}C atmospheric abundance over the last 400 years.

The production rate for the various cosmogenic isotopes can differ considerably and one can envision that many environmental parameters affect the overall production. In the following, we therefore present the averaged numbers and features for two, for our discussion the most important long-lived isotopes ^{10}Be and ^{14}C that are being produced by cosmic ray activity.

^{10}Be is mainly produced in the upper atmosphere by neutron and proton induced spallation processes on atmospheric ^{16}O and ^{14}N isotopes. With typical cross sections of $\sigma = 1.3$ mb for ^{14}N$(p, x)^{10}$Be and 0.9 mb for ^{16}O$(n, x)^{10}$Be, the production varies between $5 \cdot 10^{-5}$ atoms/g/s at 0° latitude and $2.5 \cdot 10^{-5}$ atoms/g/s at 90° latitude, at stratospheric altitude (15–50 km altitude). In the troposphere (5–15 km altitude) the secondary neutron and proton flux reaches a maximum; the production rate reaches a latitude independent value of $\approx 10^5$ atoms/g/s. The averaged global production rate is $4 \cdot 10^{-2}$ atoms/g/s.

Atmospheric ^{14}C is mainly produced by the capture of thermalized neutrons on atmospheric ^{14}N. High-energy cosmic ray protons in the stratosphere produce the neutrons by spallation processes. Most of the neutrons penetrate into deeper atmospheric layers and thermalize before they are captured by nitrogen nuclei forming ^{14}C in the ^{14}N(n,p)^{14}C reaction, as schematically shown in Figure 1.39. The cross section for the reaction is 1.8 mb. By this mechanism ^{14}C reaction is produced throughout the atmosphere, almost equally partitioned between the stratosphere and the troposphere. The production rate has its maximum in polar regions at 4 atom/cm^2/s, declining to only 1 atom/cm^2/s near the equator. This is mainly caused by the influence of the geomagnetic field focusing the primary cosmic ray particles into the pole areas. The average production rate is 2.2 atoms/cm^2/s, corresponding to a total production of approximately 7–8 kg ^{14}C per year.

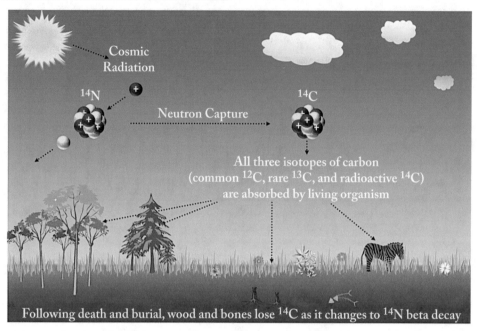

Figure 1.39: Schematic of ^{14}C production and decay in the atmosphere. ^{14}C is produced in the atmosphere by cosmic neutrons colliding with Nitrogen atoms. The newly formed ^{14}C is oxidized to ^{14}CO$_2$ where it then enters the biosphere. Following an organism's death, radioactive decay gradually converts the ^{14}C in the now dead material back to ^{14}N.

The ^{14}C atoms couple chemically with the oxygen in the atmosphere forming radioactive ^{14}CO$_2$, which takes part in the biochemical carbon cycle.

Through the biological carbon cycle, ^{14}C is absorbed into plants in amounts comparable to the fraction in the atmosphere. In this way, ^{14}C becomes part of the general food cycle and, thereby, the biological composition of animals and people. Bodies maintain the absorbed or inhaled equilibrium of carbon isotopes until death, when food and air intake stop. While the ratio of the stable carbon isotopes remains constant in the body materials, radiocarbon decays and so, the ratio of remaining radiocarbon to stable carbon provides information on the amount of time that has passed since death.

1.8.2   RADIOGENIC RADIATION

In addition to cosmogenic radiation, there is a substantial amount of radiogenic radiation in the natural environment. That is due to the decay of long-lived radioisotopes that have been produced in previous star generations, well before the origin of the solar system. This material

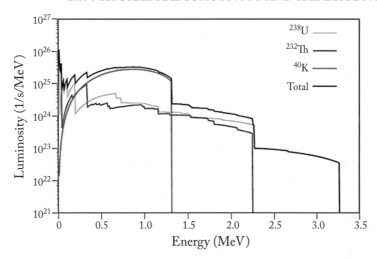

Figure 1.40: The differential geo-neutrino luminosity assuming the following global abundances (according to the BSE model): $a(^{238}\text{U})$ = 13.4 ppb, $a(^{232}\text{Th})$ = 53.6 ppb, and $a(^{40}\text{K})$ = 157.4 ppm.

was imbedded in the proto-solar cloud and deposited in solar and planetary material during formation of the early planetary system.

On earth, the heat generated by the radioactive decay of these long-lived isotopes maintains a hot Ni–Fe core surrounded by a boiling solar mantle. The amount of radioactive decay processes in the solar interior has been directly determined and confirmed through the measurement of the geothermal neutrino flux generated by the decay processes, as shown in Figure 1.40. The convective motion in the mantle drives the continental drift and leads to the formation of the relatively thin but solid earth crust at the continental boundaries. Depending on the geo-chemical history of the crust matter, it contains a fair amount of radiogenic material. In particular, granite shows a high amount of radioactivity.

The dominant long-lived radioisotopes are the actinide isotopes $^{232}\text{Th}(t_{1/2} = 1.405 \cdot 10^{10}$ y), $^{235}\text{U}(t_{1/2} = 4.468 \cdot 10^9$ y), ^{238}U, $(t_{1/2} = 4.468 \cdot 10^9$ y), $^{237}\text{Np}(t_{1/2} = 2.14 \cdot 10^6$ y), and the potassium isotope $^{40}\text{K}(t_{1/2} = 1.251 \cdot 10^9$ y). All of these isotopes provide an important contribution to the natural radioactivity level and can be used for cultural heritage studies. Regarding the decay chains, these studies are primarily based on the characteristic abundance distribution in the decay chain and the associated fractionation chemistry in the materials.

For ^{40}K, we don't have a decay chain but the important aspect to consider is the ratio of mother and daughter nucleus ^{40}Ar, which has emerged as a very promising dating tool— the so-called argon-potassium method. Radioactive ^{40}K exists in large quantities and its natural abundance is 0.012% (120 ppm) of the total amount of potassium found in nature. This represents quite a large number of potassium nuclei in nature. The decay mechanism is rather unique since ^{40}K is a rare example of an isotope that undergoes all three types of weak interaction, driven by β^-, β^+, and electron capture decay. About 89.28% of the time, ^{40}K decays to ^{40}Ca with emission of a β^- particle or an electron and an antineutrino. About 10.72% of the time, ^{40}K decays to ^{40}Ar by electron capture, with the emission of a 1.460 MeV gamma ray and a neutrino. Only 0.001% of the time will ^{40}K decay to ^{40}Ar by emitting a β^+ or a positron and a neutrino. The radioactive decay of this particular isotope explains the large abundance of argon (nearly 1%), which escaped as decay product into the earth's atmosphere.

The uranium and thorium actinides decay, through a long chain of α- and β-decay processes, toward stable lead material. The characteristic decay sequences for each of the long-lived actinides are shown in Figures 1.41 and 1.42. The emitted particles are being stopped in the surrounding environment and the kinetic energy from the decay process is converted into thermal energy or heat. Neutrino measurements indicate that about 50% of the internal energy of earth is due to radioactive decay, while the rest is coming from the release of gravitational energy originated during the planetary formation process.

While the emitted α and β radiation is typically quickly absorbed in the surrounding rock and ground environment and the initial kinetic energy is distributed by multiple elastic scattering events, the α particles with typical energies of 3 to 7 MeV can also induce nuclear reactions, most notably (α, n) reactions on light material such as ^{11}B and ^{13}C. These reactions translate into an appreciable neutron flux, which can further alter the isotopic composition of the material via secondary neutron capture reactions. The actinide amount in natural materials depends on the geo- and/or biochemical history. There is a large variation of actinide composition in the different rock formations and ground materials. These kinds of materials have been used for everything from making tools to constructing dwellings, throughout human history. Identifying specific radioactive features can provide a unique analytical tool for determining the age of artifacts and for identifying and characterizing early fabrication techniques and transport patterns, as will be discussed later.

Figures 1.41 and 1.42 show these decay chains with the detailed decay branches, which are due to parallel occurring decay processes. Figure 1.41 shows the sequence of decay processes originated by the long-lived ^{232}Th and ^{237}Np isotopes feeding a number of shorter-lived isotopes and leading to the stable isotope. Figure 1.42 displays the feeding patterns of stable ^{207}Pb and ^{206}Pb by the decay of ^{235}U and ^{238}U, respectively.

Within the $4.6 \cdot 10^9$ years since the formation of earth, an appreciable amount of the initial long-lived isotopes has decayed and the overall radioactivity level has declined, as shown in Figure 1.43, but the effects of the decay still have a substantial impact on the general radiation exposure in the Holocene. The abundance of long-lived actinides in the ground continue to generate an appreciable radiation background in the environment that, next to the cosmic ray induced background, can be utilized in many applications for cultural heritage studies.

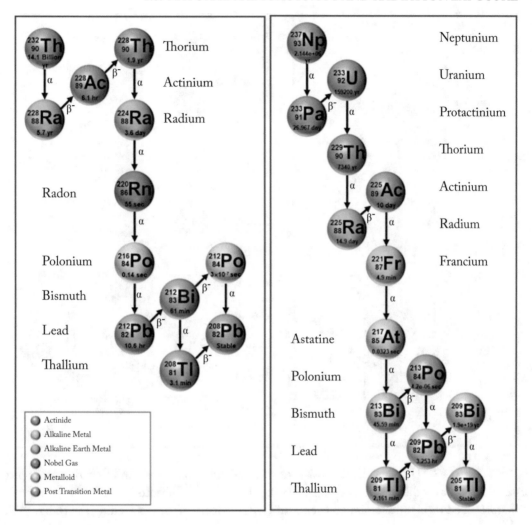

Figure 1.41: The ^{232}Th decay chain consists of five α-emission events and four β^--decay processes feeding the stable ^{206}Pb isotope. The ^{237}Np decay chain is characterized by eight α-emission events and four β^--decay processes feeding the stable ^{205}Tl isotope.

1.8.3 ANTHROPOGENIC RADIATION

There is a third radioactivity component that adds to the average exposure to radiation; this is the anthropogenic radiation, which is due to human activity. This component has grown considerably over the last few decades and now statistically represents about 50% of the average human radiation exposure. There are multiple sources for anthropogenic radiation and these include: radiation emission from nuclear power plants, in particular from nuclear power plant incidents;

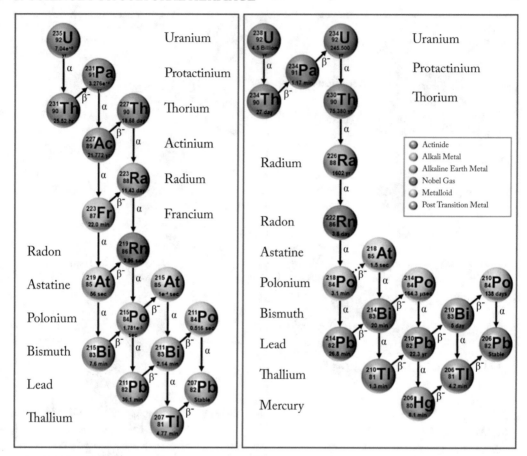

Figure 1.42: The ^{235}U decay chain consists of seven α-emission events and four β^--decay processes that occur via different branchings, feeding the stable ^{207}Pb isotope. The ^{238}U decay chain is characterized by eight α-emission events and six β^--decay processes feeding the stable ^{206}Pb lead isotope.

technically enhanced radioactivity—which represents mining, coal and fossil fuel burning—by which long-lived radioactivity is being transported from inaccessible regions underground to be released into the environment and the atmosphere. In particular, anthropogenic radiation is dominated by the growing exposure to radiation due to medical diagnostic and treatment procedures.

Even the experimental testing of cultural heritage materials provides a source of anthropogenic radiation. First there is the X-ray of neutron radiography techniques, by which hidden layers of materials in paintings and other artifacts can be made visible; these techniques use ra-

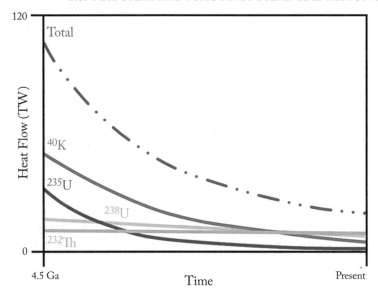

Figure 1.43: The decay curve for the natural long-lived actinide radioactivity in earth material. The original activity, dominated by the contributions of the ^{235}U chain and by ^{40}K, has declined to about 20% of its initial value. Presently, the main contributions are coming from the decay processes associated with the ^{238}U and the ^{232}Th chains.

diation from X-rays, muon, and neutrons produced in X-ray generators, synchrotron, and other accelerators, respectively, as well as nuclear reactors to generate the necessary radiation flux.

In addition, and of particular interest for cultural heritage applications, anthropogenic radiation includes the fall-out from weapons test programs in the 1950s–1960s. The so-called nuclear bomb peak, see Figure 1.44, has emerged as a highly sensitive tool for distinguishing between forgery and original, in the study of potentially valuable historical artifacts. During the nuclear test program, appreciable amounts of long-lived radiation were ejected into the higher troposphere and stratosphere atmosphere, distributed globally by high altitude winds, and deposited by fall-out to the ground.

This, in particular, is the case for ^{14}C, which was produced by high neutron flux generated by the fission process, via the ^{14}N(n,p)^{14}C reaction on atmospheric nitrogen. Chemically, this process forms radioactive carbon dioxide ^{14}CO$_2$, which like any other carbon dioxide is absorbed by plant material in the carbon cycle and deposited into the biological food cycle. This means that with the bomb test program, all biological life carries a larger ^{14}C radioactivity level than during earlier times. While this is not dangerous due to the very low energy of the β decay of ^{14}C, it re-sets the clock for radiocarbon dating. Newly produced materials show a significantly larger ^{14}C content than old materials, a fact that is utilized today, in particular for the deter-

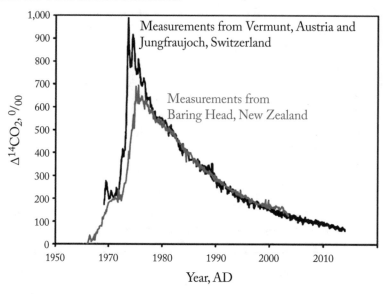

Figure 1.44: The increase of the ^{14}C abundance, chemically bound as $^{14}CO_2$ in the northern and southern hemisphere during and after the nuclear bomb test program. While the half-life of ^{14}C is 5730 years, the observed curve shows a time of only 15–20 years for the peak to decline to pre-testing conditions. This is due to the rapid processing of the $^{14}CO_2$ by the natural carbon cycle. The ^{14}C is being imbedded into all of the earth's biological material.

mination of modern art forgeries. In this context, it should also be pointed out that the rapid increase in the burning of fossil fuel leads to an inverse effect. Fossil fuel comes from millions of years-old deposits of biological materials; its initial ^{14}C has decayed and is gone. While burning fossil fuel gradually increases the CO_2 level in the atmosphere, causing heat trapping and global warming, it has emitted $^{14}CO_2$ depleted carbon dioxide into the atmosphere, causing an overall decline in the atmospheric ^{14}C related radioactivity since the 1830s. However, this decline is now overshadowed by the bomb peak.

Besides the light ion fall-out products of nuclear testing, such as ^{14}C but also tritium that were produced by neutron-induced reactions in the atmosphere, long-lived radioactive fission products, most notably ^{90}Sr and ^{137}Cs, were also deposited into the atmosphere and became part of the radioactive fall-out. The half-life of these isotopes is approximately 30 years; this means that more than 50% of the initially deposited activity has decayed away to stable ^{90}Zr and ^{137}Ba, respectively. Radioactivity with a shorter half-life has decayed long before to stable isotopes (Figure 1.45).

Radioactive fall-out has become useful for art forensics applications, particularly for the identification of modern forgeries produced during or after the test program, since any oil paint-

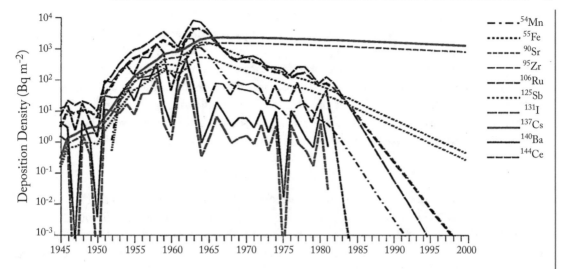

Figure 1.45: Worldwide cumulative deposition density of radionuclides produced in atmospheric testing. The monthly determined results have been averaged over each year. The figure only includes longer-lived radioisotopes.

ing made in the nuclear era will show traces of Caesium-137 and Strontium-90. These radioactive impurities are nonexistent in paintings made during the pre-nuclear era. Also, carbon-based materials produced during or after the bomb test period show a higher abundance of 14C than older materials. This is now an important tool in AMS-based applications.

Additional radioactive components are inserted into the environment by reactor accidents, such as Chernobyl and Fukushima; the fall-out is, however, mostly confined locally since the material is not being injected into the upper troposphere and therefore not subject to distribution by high altitude winds.

1.9   ACCELERATORS AND REACTORS AS SOURCES FOR ANTHROPOGENIC RADIATION

It was naturally desirable to devise methods by which one would to be able to produce strong beams of atomic projectiles simply by accelerating various charged particles in strong electric fields.

(George Gamow, Atomic Energy in Cosmic and Human Life, 1947)

Many of the analytical techniques used in the study of cultural heritage objects are based on the production of radiation for probing the different samples. The primary tools are particle accelerators, which are used to accelerate charged particles to desired energies for spectroscopy analysis such as PIXE or PIGE. Also, mass separation and accelerator mass spectrometry are based on the use of particle acceleration techniques. Accelerators are also being used for generating intense gamma or neutron sources for activation techniques. While accelerators are universal tools, small research reactors are being used primarily for neutron activation techniques. Specific samples are positioned in the neutron flux, near the reactor core, to measure the specific activity of the neutron capture products and the associated time dependence of the decay in order to learn something about the composition of the artifact. In the following section we provide a short summary of the accelerator types that are presently used in cultural heritage studies and a short description of reactor-based analytical methods.

1.9.1   ELECTROSTATIC ACCELERATORS

Electrostatic accelerators are based on the principle that charged particles can be accelerated in a static electric field. An electric field exists between two different electrical potentials, with a potential difference of U. A charged particle with charge q is accelerated to a kinetic energy $E_{kin} = qU$. To reach kinetic energies necessary for material studies using atomic or nuclear physics techniques, the energy must typically be in the range between 1 and 10 MeV. The Electronvolt is the kinetic energy that a single charged particle picks up in an electrical potential of 1 V. This means the required potential necessary for electrostatic accelerators is measured in Millions of Volt. These kinds of potentials are typically produced by transferring charge from a grounded pole to an electrically isolated pole, the so-called terminal (Figure 1.46). The charge can be transported via isolated rubber belts (the Van de Graaff system), or via chains of insulated metal pellets (the Pelletron system). The potential corresponds to the amount of charge Q that can be transferred to the terminal and the specific capacitance C of the system

$$V = Q/C. \tag{1.77}$$

The amount of charge depends on how much charge the carrier can transport, which in turn depends on the quality of the belt or the Pelletron chain. In a typical electrostatic accelerator, up to 10 MV potential difference can be achieved. This translates into a particle energy of $E = Q \cdot V$. For a 10 MV machine, a positive charged proton will be accelerated to $E = 10$ MeV.

The such-achieved terminal potential V is maintained by an equilibrium between the up-charge current I_Q, the current through the resistor chain I_R, the discharge current to the corona points I_C that prevents overcharging of the terminal, and the beam current I_B, following Kirchhoff's current law

$$I_Q = I_R + I_C + I_B. \tag{1.78}$$

The beam intensity I_B is directly correlated with the charging capabilities of the charge carrier. A single-ended Van de Graaff can produce high-intensity beams but is limited to relatively low voltage, due to the up-charge capabilities of the charging system. The beam is generated in a so-called ion source that produces a plasma of ionized particles—typically by radiofrequency technique. The positively charged particles are extracted out of the source and accelerated in the electric acceleration field toward the target. To ensure homogeneous acceleration, the potential

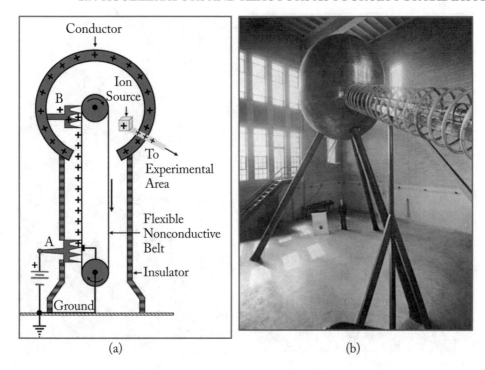

Figure 1.46: The Van de Graaff principle of charging an electrically isolated sphere (terminal) over a rubber band or any other charge carrier, building up a potential difference. Charged particles are being produced at the terminal and accelerated toward the ground level (a) and the 1937 model of the 1 MV open-air accelerator at the University of Notre Dame (b). The charging belt comes from the right toward the terminal sphere and the acceleration tube leads toward the front. The potential gradient is maintained by external guard rings.

gradient within the acceleration tube is kept constant by a series of acceleration sections, which are coupled through a resistor chain.

High voltages in excess of 1 MV cannot, however, be achieved in open air. Violent discharges occur, particularly in high humidity environments (Figure 1.47a). The maximum potential difference in open air is around 1 MV. For this reason, electrostatic accelerators reaching higher voltages are mounted in pressurized tanks containing a dry CO_2 (carbon dioxide) and N_2 (nitrogen) gas mixture or, in more modern machines, of SF_6 (sulfur hexafluoride) gas (Figure 1.47b). With the right gas mixture, discharges in the tank can be minimized and higher voltages can be achieved.

Van de Graaff and Pelletron accelerators in the 2–5 MeV range are well suited for PIXE or PIGE material studies. In these accelerators, the proton beams are generated in an RF plasma source and accelerated to the desired energies. The beam intensity needs to be limited to a few

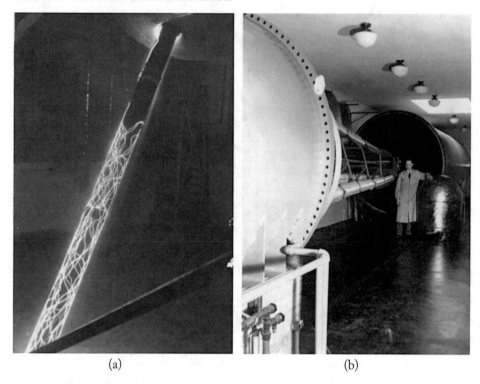

(a)   (b)

Figure 1.47: Discharge along the insulating support legs of the 1937 open-air Notre Dame accelerator (a). Pressurized tank for the second-generation 1942 Notre Dame accelerator that reached a terminal voltage of 4 MV (b).

nano-Amperes—well below the capabilities of such machines—to avoid damage of the materials [1]. For Accelerator Mass Spectrometry (AMS), other types of electrostatic accelerators are being used, the so-called tandem accelerators, which can easily accelerate heavy ions. While PIXE relies on proton (or other charged particle) beams bombarding the materials to be analyzed, AMS relies on generating a beam of isotopes from the material itself and accelerating it. This requires different ion sources and acceleration toward higher energies.

To generate higher voltages on the basis of the electrostatic acceleration principle in the 1960s, the tandem accelerator was invented [2]. The tandem consists of two acceleration sections with a positively charged high voltage terminal in between (Figure 1.48). The potential V at the terminal was generated by belt-driven up-charge, as described above. The ion source was on ground level and produced negative ions by charge pick-up techniques. The negative ions were accelerated toward the high voltage terminal, where the electrons were removed in a stripper foil or gas. The now positively charged ions entered the second acceleration tube and were accelerated toward ground. The total energy was $E = (q + 1) \cdot V$ with q being the charge

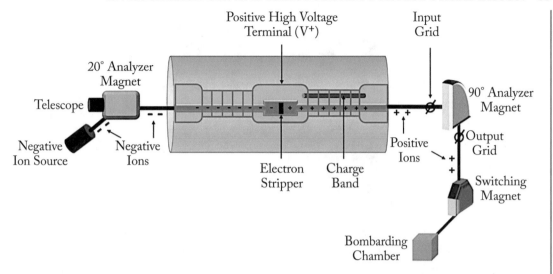

Figure 1.48: Principle of a tandem accelerator, where a negatively charged beam is injected into the acceleration tube and accelerated toward the positively charged terminal. A stripper system—either a thin carbon foil or a gas—rips the electrons away from the impinging isotopes, turning a negative beam into a positive beam with a certain charge state distribution. One of the charge states is being selected by the analyzing magnet for further experimental use.

state of the positive ions selected by the analyzing magnet mounted behind the accelerator. This design of accelerating a negative beam toward a positively charged terminal, and then stripping the ions to some positive charge distribution allowed for gain of much higher energies than possible hitherto. The first tandem developed was the EN tandem by HVEC with 5.0 MV terminal voltage, followed later by the development of the FN tandem that could reach 7.5 MV terminal voltage [3].

The problem with the tandem machines was that they required negatively charged ions for the first stage of operation; for several elements, negative ions were difficult or impossible to produce [4]. This includes all noble gases with closed electron shells. The only observed exception to this requirement of closed electron shells is the nitrogen atom, which has a half-filled outer shell configuration ($1^s2 2s^2 2p^3$) that cannot be formed since this would violate the Pauli Exclusion Principle [5]. The production of these isotopes requires special means and techniques in both accelerator as well as ion source technology.

The AMS approach is based on the analysis and counting of long-lived radioactive isotopes with well-known half-lives, such as ^{14}C or ^{10}Be, from the material to be investigated. This requires the use of sputter sources, where the prepared materials in a pellet are being bombarded by a cesium beam, sputtering the pellet material away. The sputtered isotopes pick up electrons

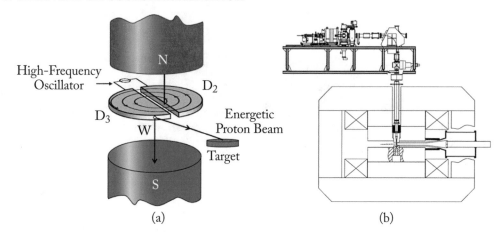

Figure 1.49: Basic design scheme of a cyclotron accelerator with the two Dees separated by a gap, across which the particles are accelerated by an alternating voltage each time they reach the gap on their orbital trajectory in the magnetic field (a). The external ion source producing the beam particles and injecting them into the center of the device (b).

from the cesium and are extracted by electrical fields, to be injected into the acceleration tube. Further details on the actual AMS method will be discussed in a later chapter.

1.9.2   CYCLOTRONS

Cyclotron accelerators were invented in the 1930s by Ernest Lawrence and collaborators. These accelerators are based on a pulsed acceleration system with particles crossing, alternating voltage gaps along a circular orbit. The core of the cyclotron consists of two hollow semicircular electrodes, called Dees, mounted back to back, separated by a narrow gap, in an evacuated chamber between the poles of a magnet. An alternating electric field is created across the gap by a radio-frequency oscillator. The beam particles are produced from an external ion source and injected near the center of the cyclotron in the gap, where they are accelerated in the electric field across the gap (Figure 1.49).

The perpendicular magnetic field guides the moving particle into a circular trajectory according to the Lorentz force. The time T for completing a full orbit depends on the mass m and charge q of the particle and the strength of the magnetic field B

$$T = \frac{2\pi m}{qB}. \tag{1.79}$$

By the time the particles have completed a semi-orbit and return to the gap, the electric field has reversed, so they are accelerated again. Due to this acceleration pulse, the speed of the particles and the radius of their orbit increase each time they cross the gap, while the circulation

(a)   (b)

Figure 1.50: The 60-inch cyclotron (the 60 inches refer to the diamete of the magnet poles) at the University of California in 1939 (a) and a modern commercial cyclotron developed by the Belgium company IBA for the production of medical isotopes (b).

time remains constant as long as the mass of the particles and the strength of the magnetic field remain constant. Therefore, the cyclotron can be operated at a constant frequency for the alternating voltage, the so-called cyclotron frequency:

$$\omega = \frac{q}{m} \cdot B. \tag{1.80}$$

A cyclotron operating in this manner can accelerate protons to energies no greater than 25 million electron volts. At these energies the effects of special relativity become a very important factor. Approaching the speed of light causes a relativistic increase in mass for the accelerated particles. As the mass increases, the orbital frequency decreases, and the particles crossing the gap are out of frequency with the alternating voltage, causing the electric field to decelerate them. To overcome this limit, the isochronous cyclotron was developed, which relies on local variations of the guiding magnetic field, adapting for the increasing relativistic mass of particles during acceleration (http://hyperphysics.phy-astr.gsu.edu/hbase/magnetic/cyclot.html).

Today, cyclotron accelerators (Figure 1.50) have a wide range of application, from commercial use for medical isotope production to fundamental research in nuclear physics and material sciences. They are less used in cultural heritage studies, but they certainly would be suitable for these kinds of applications [6].

1.9.3   SYNCHROTRON RADIATION

In the last two decades, the use of small accelerators for cultural heritage studies has been complemented by the increasing use of so-called synchrotron light sources. Light sources are based on the emission of electromagnetic radiation by accelerated electrical particles. Synchrotron radiation is emitted by electrons circulating in a storage ring at nearly the speed of light (Figure 1.51). The synchrotron radiation covers a broad spectral range from X-ray, ultraviolet, visible to infrared range. The electrons need to be accelerated to the desired energies before being injected into the storage ring. This is usually done by a small linear accelerator (linac) system. In the storage ring the particle runs, unlike in a cyclotron, on a stationary circular path, which is maintained by bending the electron trajectory by use of magnetic fields, causing the electrons to emit photons and thus producing the synchrotron radiation.

To accommodate for the associated energy loss, the electrons are being continuously re-accelerated by a high-frequency alternating electric field (Figure 1.51). The adaptation for the relativistic velocity of the particles is done by variation of the magnetic field strength in time, rather than in space as in the isochronous cyclotron, ensuring the constant radius of the trajectory. For particles that are not close to the speed of light, the frequency of the applied electromagnetic field may also change to follow their non-constant circulation time. This allows the vacuum chamber for the particles to be a large thin torus, rather than a disk as in the cyclotron design. Also, the thin profile of the vacuum chamber allows for a more efficient use of magnetic fields than in a cyclotron, enabling the cost-effective construction of larger synchrotrons. Electron beams of up to several Amperes with energies up to GeV can be obtained, causing very intense light emission.

To emit enhanced radiation, a wiggler system, consisting of several magnets, is used to periodically deflect (wiggle) the electron beam inside the storage ring. These deflections create a change in acceleration, which in turn produces tangential emission of broad synchrotron radiation with enhanced intensity for the desired application [7].

The extremely bright synchrotron light is extracted at certain bending points of the storage ring. Using prisms or grating techniques, a monochromatic light beam is selected and focused through a mirror system onto the sample, causing the emission of a characteristic X-ray spectrum (Figure 1.52). In this way, a range of materials—from objects of atomic and molecular size to human-made materials with unusual properties—can be investigated. Increasingly, the synchrotron radiation technique is also used for the analysis of cultural heritage objects as will be discussed later [8].

1.9.4   NUCLEAR REACTORS

The main purpose of nuclear reactors in cultural heritage studies is to utilize the high neutron flux for radiating anthropological, archaeological, or historical samples. The neutron radiation turns stable isotopes in the sample into radioactive species. Through analysis of the specific radioactive

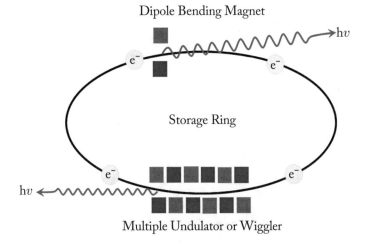

Figure 1.51: A simplified representation of the synchrotron radiation source. The electron storage ring is composed of many straight sections, connected by the dipole bending magnets keeping the electrons on an orbital trajectory. An undulator or wiggler is installed along the straight portions of the storage ring to generate an intense beam of synchrotron radiation for specific applications.

decay, one can deduce the isotopic and elemental composition of the material, which provides information about its provenance and possibly also manufacturing process.

The neutron flux inside the reactor core is of up to 10^{12}–10^{15} neutrons/s cm^{-2} or when transferred by a neutron ladder or a beamline outside the reactor yields 10^{10}–10^{12} neutrons/s cm^{-2}. The intensity of the neutron flux depends on the reactor type and the actual positioning of the sample and needs to be calibrated by radiating a well-known calibration sample, to which the results can be normalized.

Neutron production in nuclear reactors is based on the principles of nuclear fission. Uranium and plutonium isotopes are subject to neutron induced fission under release of energy and additional neutrons. ^{235}U and ^{239}Pu provide the typical reactor fuel. The fuel rods are enriched to a few percent of ^{235}U or ^{239}Pu, depending on the reactor type; the fission of these isotopes generates the neutron flux in the reactor core. These neutrons are used for causing additional fission processes, but their flux is limited by so-called control boron or cadmium-loaded control rods, which absorb most of the neutrons. The positioning of the control rods therefore maintains the orderly reactor operation. The released radiation heat is transported away through a cooling water cycle, preventing overheating of the reactor system (Figure 1.53).

For research reactors, the neutrons can be used for a large range of applications, particularly also in the material sciences. The applications for cultural heritage analysis [9] have become an important aspect for the nuclear reactor research community [10].

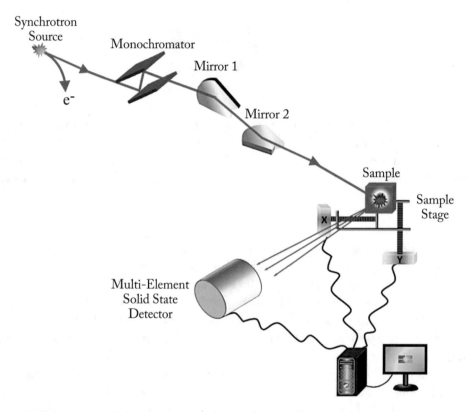

Figure 1.52: Schematics of the endstation of a synchrotron light experiment.

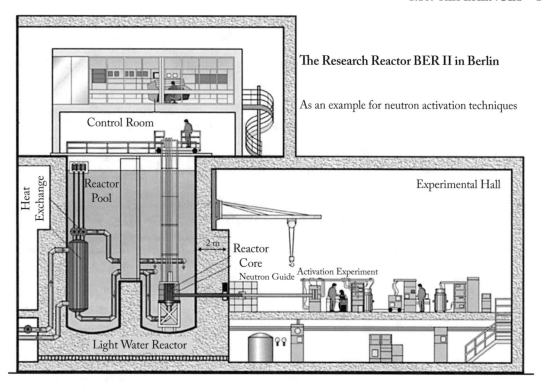

Figure 1.53: The design of the light water research reactor (BER-II in Berlin). Samples can be either placed near the reactor core where they can be exposed to a maximum neutron flux, or the neutrons can be guided along a beamline from the reactor core to an experimental station to activate a sample at a lower flux.

1.10 REFERENCES

[1] P. Milota, I. Reiche, A. Duval, O. Forstner, E. Guicharnaud, W. Kutschera, S. Merchel, A. Priller, M. Schreiner, P. Steier, E. Thobois, A. Wallner, B. Wüenschek, and R. Golser, PIXE measurements of renaissance silverpoint drawings at VERA, *Nucl. Instr. Meth. B*, 266:2279–2285, 2008. DOI: 10.1016/j.nimb.2008.03.005. 80

[2] R. J. Van de Graaff, Tandem electrostatic accelerators, *Nucl. Instr. Meth.*, 8:195–202, 1960. DOI: 10.1016/s0029-554x(60)80006-7. 80

[3] J. G. Trump, New developments in high voltage technology, *IEEE Transactions on Nucl. Science*, 14:113–119, 1967. DOI: 10.1109/tns.1967.4324534. 81

[4] R. Middleton, A negative ion cookbook, *BNL TVD*, 1990. www.pelletron.com/cookbook.pdf 81

[5] W. P. Wijesundera and F. A. Parpia, Negative ions of carbon, nitrogen, and phosphorus, *Phys. Rev. A*, 57:3462–3468, 1998. DOI: 10.1103/physreva.57.3462. 81

[6] L. Beck, Recent trends in IBA for cultural heritage studies, *Nucl. Instr. Meth. B*, 332:439–444, 2014. DOI: 10.1016/j.nimb.2014.02.113. 83

[7] H. Winick, G. Brown, K. Halbach, and J. Harris, Synchrotron radiation wiggler and undulator magnets, *Phys. Today*, 34:50–63, 1981. DOI: 10.1063/1.2914568. 84

[8] D. Creagh and D. Bradley, Physical techniques in the study of art, *Archaeol. and Cult. Herit.*, 2:1–95, 2007. DOI: 10.1016/s1871-1731(07)x8002-3. 84

[9] N. Kardjilov and G. Festa (Eds.), Neutron methods for archaeology and cultural heritage, *Springer Nature*, Switzerland, 2017. DOI: 10.1007/978-3-319-33163-8. 85

[10] Nuclear techniques for cultural heritage research, *IAEA Radiation Technology Series*, no. 2, Ed., M. Haji-Saeid, Vienna, 2011. https://www-ub.iaea.org/MTCD/Publications/PDF/p1501_web.pdf 85

CHAPTER 2

Spectroscopy

2.1 INTRODUCTION

When the sun rose a few spans, he again closed the windows to the beautiful world outside with everything that lived and weaved and let just a single ray of light into the darkened room, through a small hole he had drilled in the shutter. When this ray was carefully stretched to the ordeal, Reinhart began his day's work without further hesitation, picked up paper and pencil to continue where he had paused the day before.

(Gottfried Keller, Das Sinngedicht, 1855)

Spectroscopy techniques investigate the interaction between electromagnetic radiation and matter. Spectroscopy emerged during the 19th century with the use of prisms to separate light into its different wavelength components or colors. The technique opened the window to study the composition of stellar light and identify the characteristic radiation emanating from gaseous materials. This led to the realization that each element emits its characteristic spectrum when excited through energy transfer. The concept was expanded to include interaction with any electromagnetic radiation, as a function of wavelength (frequency). Spectroscopic data is often represented by an emission spectrum, a plot of the response of interest as a function of wavelength or frequency. There are four main groupings of methods in optical spectroscopy. These are absorption, emission, scattering, and luminescence. There are a few other methods such as reflection, refraction, diffraction, and polarization that are also used for observing the interaction of radiation with matter. Atomic spectroscopy investigates absorbed and emitted electromagnetic radiation by atoms. As each element has unique characteristic spectra, these spectroscopy methods are used for the determination of elemental compositions. This chapter discusses atomic and optical (vibrational) spectroscopic methods that employ X-ray, UV, visible, and IR lights to investigate the structure and composition of materials in cultural heritage objects.

2.2 GENERATION OF X-RAYS

X-rays are electromagnetic radiation in wavelengths that range from 0.01–10 nm. X-ray radiation is produced in generators or cyclic particle accelerators (synchrotrons). In X-ray generators, a target material incorporated into a high-vacuum tube is bombarded by energetic particles (Figure 2.1). In the most common X-ray tubes, electrons emitted from a source (filament) accelerate due to a significant electrical potential that is applied between the source and the target (usually made of Cu, Mo, or Cr). These electrons bombard the target and a part of their kinetic

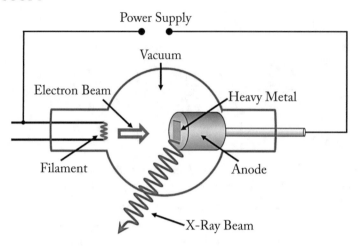

Figure 2.1: Production of X-rays.

energy emits as X-ray radiation. Such bombardment also generates photoelectrons and Auger electrons, as well as heating the metallic anode material. Two types of X-rays are emitted in this process: a continuous *white radiation*, also called Bremsstrahlung, and a series of sharp lines that are specific to the target material.

Bremsstrahlung radiation is generated due to the conversion of the kinetic energy of electrons, upon their collision with the target atoms. The maximum energy of the X-ray photon is given by:

$$E_{max} = h\nu_{max} = \frac{hc}{\lambda_{min}} = eV, \qquad (2.1)$$

where ν_{max} is the largest frequency, h is Plank's constant, e is the charge of the electron, V is the applied voltage between filament and target, c is the speed of the light, and λ_{min} the minimum wavelength as determined by

$$\lambda_{min} = \frac{hc}{eV} = \frac{12398}{V}. \qquad (2.2)$$

This equation indicates that the higher the voltage applied, the shorter (more energetic) the minimum wavelength (Figure 2.2a). The kinetic energy of most electrons does not entirely transform to X-ray photons during a single collision. Incoming electrons hit inner shell electrons in the target atoms, and a portion of their energy emits as X-ray photons. Therefore, the spectrum of Bremsstrahlung radiation has a minimum wavelength and continues to longer wavelengths, until all of the kinetic energy is absorbed. The incoming electrons also collide with other electrons in the target, emitting lower energy X-ray photons. The collision of incoming electrons with outer shell (valence) electrons of the target produces heat.

Increasing the energy of the incident electron beam above a certain threshold causes the emission of additional intense X-ray lines (Figure 2.2b). The energies of such X-ray lines are

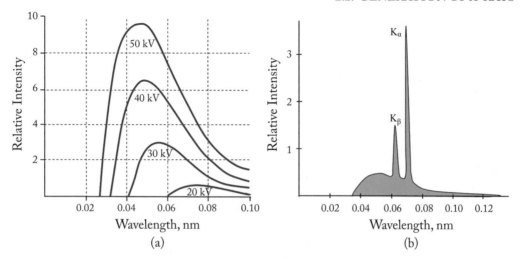

Figure 2.2: Bremsstrahlung radiation (a) and characteristic X-rays (b).

specific to the target material. For example, the energy transferred from an incoming electron to a metal atoms causes the removal of the core level electron. This can occur only when the kinetic energy of the incoming electron exceeds the binding energy of the electron within the target atom. In the second stage, an electron in a higher energy state moves to the vacant hole in the target atom. Such electronic transition is accompanied by the emitting of an X-ray photon. These X-rays are known as characteristic radiation, since the energy of the photon is characteristic of the type of atom used for the anode material, as outlined in Section 1.2. For example, an electron removed from the tungsten K-shell with a binding energy of 69.5 keV creates a vacancy, which is filled by an electron from the L shell with a binding energy of 10.2 keV. The X-ray photon emitted during such a transition is equal to $69.5 - 10.2 = 59.3$ keV.

The same anode material usually emits several characteristic X-ray energies due to the presence of different energy levels (such as K, L, M). Electrons from the incoming electrons can, however, remove these levels, and the vacancies can be filled from different energy levels of the anode material. Figure 2.3 shows the energy levels in tungsten and some of the electronic transitions that result in characteristic X-rays. Each characteristic X-ray is assigned by the shell name in which the vacancy occurred. To identify the transition, a subscript showing the origin of the filling electron is also added. For example, the subscript α denotes filling with an L-shell electron and β indicates the electronic transition from either the M or N shell.

For practical applications such as artifact evaluation, monochromatic X-rays have greater importance. Therefore, X-ray tube generators are inefficient, as they produce white radiation and characteristic lines, Auger, and photo-electrons. However, using different metallic filters can produce monochromatic X-ray beams. Synchrotron radiation sources are capable of producing

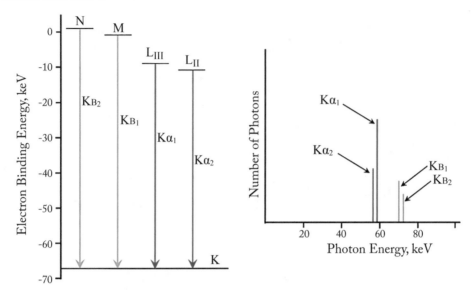

Figure 2.3: Some energy levels in the tungsten atom and electronic transitions that result in characteristic X-rays.

very intense and monochromatic X-rays with tunable wavelengths. In synchrotrons, electrons are generated by an electron source, which is being accelerated by a linear accelerator. The electrons are then transferred to a ring where they are accelerated to 3–6 GeV energies. In these energies, the speed of the electrons is approaching the speed of light. These electrons are then directed into the central ring, where they emit X-ray during circulation.

2.3   X-RAY EMISSION SPECTROSCOPIC METHODS

X-ray spectroscopic methods are based on the atomic properties of matter, as discussed in Chapter 1. Electrons in each atom have a characteristic distribution around the nucleus, and when the atom is irradiated with X-rays or accelerated particles such as protons, an electron located in the inner shells of an object's atoms can be ejected. This process results in a vacancy, and an electron from the outer shells moves into the vacancy. This electronic transition accompanies the release of a new X-ray photon. X-ray emission spectroscopic methods are based on the measurement of the energy of these newly formed X-ray photons. The characteristic distribution of electrons in each element is unique. Therefore, each element has a distinct set of electronic transitions and X-ray emission line patterns as outlined in Section 1.2.

X-ray emission spectroscopic techniques include *X-ray fluorescence* (XRF), *particle-induced X-ray emission* (PIXE), and *electron energy-dispersive X-ray spectroscopy* (EDX). In these methods, detectors are used to analyze the energies of the X-ray photons emitted from the object. The energy of the X-rays increases as a function of atomic number and so, the energy of a peak in the X-ray spectrum is specific to a particular element. The intensity of these lines is proportional to the number of emitting atoms. The pattern of the emission lines obtained from an object containing multiple elements is, therefore, the combination of the characteristic emission line patterns of each element. These features allow for identifying, as well as determining the quantity of each element. Determining the elemental composition of an object involves the use of standards and reference materials, in order to establish the calibration curve. Excitation beams can be scanned across the surface of the object to produce elemental distribution maps.

Energy dispersive analysis is the most common method of detecting X-ray emissions. The X-rays emitted from the object are focused on a detector, which often contains a 3–5-mm thick silicon Si(Li) crystal diode, as discussed in Section 1.7.3. The incoming X-ray photons ionize a large number of atoms in the detector and form electron-hole pairs. The amount of charge generated during these events is proportional to the energy of the X-ray photon. The bias applied across the detector helps to collect the released charge on a capacitor, and the resulting voltage pulse is amplified by a preamplifier. The output signal of the preamplifier is fed to the main amplifier, and the pulses are converted to a digital signal and sent to an analyzer. Counting the number of X-ray pulses and sorting them by their energy creates a spectrum. X-ray detectors have different resolutions and detecting speeds. The detectors should be cooled by liquid nitrogen or by a Peltier cooling system to obtain sufficiently good resolution. Energy dispersive analysis allows for rapid identification of, and quantity determination for, most elements. Wavelength-dispersive detectors are also used in the analysis of X-rays. These systems use crystals and, usually, a single wavelength is counted at a time. Therefore, multiple analyses are needed in order to complete the determination of the different elements in an object. This detection system does, however, provide better energy resolution when compared with energy dispersive detectors.

2.4   X-RAY FLUORESCENCE (XRF)

XRF is based on the excitation of atomic elements in a material component through intense X-ray radiation observing the characteristic X-ray radiation emitted by the material. XRF is the most common method used for cultural heritage investigation due to its quick and non-destructive nature, as well as its ability to detect and analyze most of the elements.

2.4.1   GENERAL CONSIDERATIONS

XRF provides qualitative and quantitative analyses of elements and, in general, is applicable for detecting all elements except the first two (H and He) of the periodic system. However, light elements—with an atomic number from 3–11 (Li to Na)—are challenging to measure. Generally, photon energies, between 1 and 25 keV, and excitation (tube) voltages, of less than 60 kV, are used for non-destructive measurements. A new generation of portable XRF systems is capable of in situ detecting, measuring multiple elements in cultural heritage objects without special sample preparation. Rapid non-distractive analysis can be done by positioning the analyzer close to a specific area of the object. These portable energy-dispersive XRF systems allow for elemental identification of cultural heritage objects with high accuracy within a few minutes.

The depth of measurement in XRF analysis depends on the element of interest and the matrix. When incident X-ray radiation enters an object, a significant part of the energy—of the X-ray photons—absorbs and converts into fluorescent photons of the various atoms. Some fraction of fluorescent photons make their way to the surface of the object and can be detected by the detector. The number of detected fluorescent photons depends on the composition of the object and its thickness. The following equation expresses the estimated sampling depth (d)

$$d = \frac{1}{\rho[\frac{\mu(E)}{\sin \psi_1} + \frac{\mu(K\alpha_1)}{\sin \psi_2}]},\qquad(2.3)$$

where $\mu(E)$ represents the absorption coefficients of the specimen for incident X-ray photons with E energy, $\mu(K\alpha_1)$ represents the absorption coefficients of observed $K\alpha_1$ fluorescent photons, ρ is the density, and ψ_1 and ψ_2 are the angles of the primary and fluorescent beams to the specimen surface, respectively [1].

The μ coefficients in Equation (2.3) are a function of incoming X-ray-photon energy (E) and the energy of $K\alpha_1$ fluorescent photons, as well as the composition of the sample with (j) elements:

$$\mu(E) = \sum_j c_j \mu_j(E) = c_i \mu_i(E) + c_m \mu_m(E)$$

$$\mu(K\alpha_1) = \sum_j c_j \mu_j(E)(K\alpha_1) = c_i \mu_i(K\alpha_1) + c_m \mu_m(K\alpha_1),$$

where the (i) represents the analyzed element and (m) the matrix. These equations indicate that sampling depth depends on density and absorption coefficients. Dense samples will decrease the information depth. Some elements can be analyzed based on two different fluorescent lines. For example, the $K\alpha$ and $L\alpha$ lines for cadmium are 23.17 and 3.13 keV, respectively; $L\alpha$ line has low energy and much higher absorption. Such a difference in X-ray line energies is sometimes an advantage for the analysis of layered materials.

Quantitative determination of elemental composition using the XRF method for cultural heritage objects is challenging. For example, in paintings a typical paint layer is made by blending the pigments with a binder consisting of vegetable oils, egg yolk, egg white, or other organic material. Usually, multiple layers of paint are applied on top of each other, followed by a coat of varnish to protect the painting. The total thickness of the paint layer can vary from several micrometers to a few millimeters. The varnish and binders, consisting of light elements, absorb some of the energy of the fluorescence X-ray photons from the heavier elements. The heaviest elements can be used to identify the pigment, as they usually have the highest concentration. For example, vermilion (mercury sulfide HgS) contains ~ 86 wt.% mercury and 14 wt.% sulfur. In a typical paint layer, mercury concentration is changing at 60–80 wt.%, whereas the sulfur amount

Figure 2.4: Analysis of pigments by a portable XRF.

is 10–15 wt%. Precise determination of the geometrical thickness and elemental composition of such a layer is difficult due to the broad range of blending ratios between pigment and binder, as well as the changing densities of pigments and binder media https://www.edp-open.org/images/stories/books/fulldl/Nuclear-physics-for-cultural-heritage.pdf.

2.4.2   EXAMPLES OF XRF

Martin de Vos (1532–1603) was an influential Flemish painter. He and his workshop created many altarpieces in Antwerp churches. Some of those altarpieces are currently in different museums, such as a triptych (painting on three panels) in the Fine Arts Museum in Seville (Figure 2.4). Portable XRF equipment was utilized for identifying pigments in paint layers of the altarpiece [2]. Lead white, $(PbCO_3)_2 \cdot Pb(OH)_2$, was identified by Pb peak and is the main component of the white color. Yellow ochre (FeOOH), as well as lead-tin yellow (Pb_2SnO_4), were used for painting yellow colors. Several red pigments—such as Fe_2O_3 and HgS, together with an unidentified organic colorant—were used for painting red areas. Smalt, a cobalt-containing pigment, was used to paint the blue areas in addition to azurite ($2\,CuCO \cdot Cu(OH)_2$).

These analyses also provided information on the white preparation layers. The presence of Ca and Pb in every XRF spectrum allowed to conclude that gypsum ($CaSO_4$) or calcium carbonate ($CaCO_3$) with white lead ($2\,PbCO_3 \cdot Pb(OH)_2$) was used for making this layer before applying other paints. In most cases, several pigments were blended or overlaid to obtain dif-

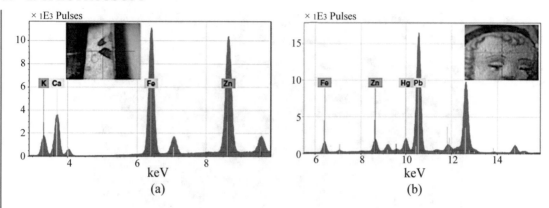

Figure 2.5: XRF spectra obtained from an illumination by Gerolamo da Cremona (a) and from Jesus' face from Virgin and Child by Franco de' Russi (b). Reprinted with permission from [3]. Copyright 2010 United States National Academy of Sciences.

ferent colors. Blending white lead with some vermilion and a copper-based green pigment, and sometimes lead-tin yellow, allowed the painter to obtain different colors. The ratio of pigments in the blend was the main factor in tuning the color and tonality. In addition to these historical pigments, the use of modern pigments such as titanium white (TiO_2), zinc white (ZnO), or chrome green (Cr_2O_3) were identified in small areas. Examination, using ultraviolet light, allowed to conclude that those were areas of intervention or retouching during restorations. Such detailed XRF results provided essential knowledge on composition that aided in restoration efforts of this masterpiece.

XRF is also widely used for the identification of elements in inks and pigments of historical manuscripts. For example, XRF analysis revealed that a colorful miniature painted by Gerolamo da Cremona (in approximately 1470, Italy) contains indigo, lazurite, azurite, malachite, and iron gall ink, lead-tin yellow, vermilion, white lead [3]. Microscopic examination revealed some shiny particles, which consist of a large amount of silver and gold, suggesting that the miniature was originally decorated with various shades of gold and silver. Another example of using XRF in the characterization of medieval manuscripts is *Virgin and Child*, illuminated by Franco de' Russi between 1460 and 1480. Microscopic examination of the various areas of the miniature revealed a large number of needle-like crystals. A needle analyzed by XRF primarily contains iron and zinc with small amounts of arsenic and mercury (Figure 2.5a). The results suggest that faces were pained with lead white, with small amounts of iron and zinc additives (Figure 2.5b). The scientists who characterized the miniature suggested that de' Russi may have deliberately added the iron gall ink in crystalline form for changing the tonality of the miniature. Medieval recipes for iron gall ink utilized $FeSO_4$ and sometimes also $CuSO_4$ mixed with gallotannic acid, a binder. These sulfate salts were contained in other metal salts such as $ZnSO_4$. Microscopic investigations suggested that, while the crystals appear on the surface, they seem to be integral parts of the paint layer and some exhibit the same cracks as the surrounding paint layer. These results suggest that XRF analysis provides significant insights, revealing recipes and techniques of old masters' paintings and manuscripts.

Modern XRF analyzers also offer poly-capillary fiber optics to produce microbeams. When using such micro-XRF systems with position sensors, a confocal XRF analysis of objects can be performed, allowing to obtain detailed elemental maps. An example can be seen in the micro-XRF investigation that characterized the pigments of a 15th century manuscript. Micro-XRF analyzer, with a ruthenium X-ray tube (50 kV and 50 W), and ultra-high intensity poly-capillary optics were utilized for elemental mapping. This system can produce a focused X-ray beam with a spot size of $\sim 30\ \mu m$.

Figure 2.6 illustrates two-dimensional macro-, millimeter-, and micron-scale elemental maps for three samples [4]. During macroscale mapping, leaves were scanned, and a total of 50,000 XRF spectra were collected. Processing of these spectra allowed for constructing maps of different metal distributions of the analyzed manuscript leaves (Figure 2.6a). The maps contain gold and calcium in gilded areas. To prepare gilded areas, it was common to apply a smooth layer of paste containing gypsum onto the parchment, and to then add a thin gold foil and polish it with a special tool. Therefore, calcium from the gypsum layer can be detected along with gold. The maximum thickness of gold foil that would allow detection of calcium in the gypsum layer can be estimated using

$$\frac{I}{I_0} = e^{-\mu \cdot x}, \tag{2.4}$$

where I is the intensity of photons from the CaKα line, with 3.67 eV energy, transmitted across the gold layer with a thickness of x, I_0 is the initial intensity of the CaKα photons, and the linear attenuation coefficient of Au is (μ). The dependence of the I/I_0 ratio, of the CaKα line on a gold layer thickness, is shown in Figure 2.7. This line indicates that the maximum thickness of gold foil that would allow the detection of calcium in the gypsum layer is approximately $\sim 1.5\ \mu m$.

Millimeter-scale mapping for the selected area showed the coexistence of Ca and Au (Figure 2.6b). Gold appears in both shiny tiles and brown lines. Shiny tiles are made of gold foils, whereas brown areas were probably prepared by a powdered gold mixed with a calcium-containing binder. Both mapping results (Figure 2.6a and b) suggest that the red pigment contains mercury and lead. The blue areas mostly consist of copper, and the brown-yellow color is associated with iron. Figure 2.6c shows a high-resolution micro-XRF elemental map. The massive green particle with a size of $\sim 250\ \mu m$ contains Cu. Two-dimensional XRF analysis provides element-specific images of flat painted surfaces. Such mapping also allows identification of the small amounts of impurities that are limited to specific areas of paintings. Knowledge on the date and the geographical origin of the object helps to limit the number of possible pigments. To fully utilize interpretive maps, extensive knowledge of XRF theory and instrumentation is needed. It should be noted that XRF mapping alone is not sufficient to identify the chemical

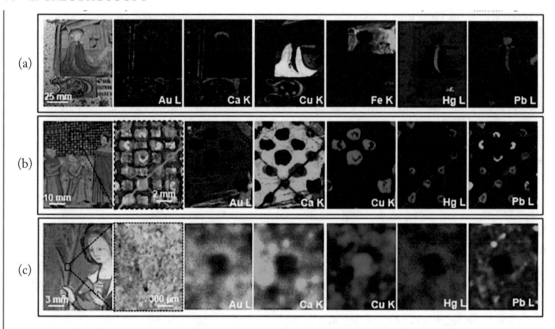

Figure 2.6: XRF elemental distribution in macro- (a), milli- (b), and micrometer- (c) scale mappings of leaves. Reprinted with permission from [4]. Copyright 2016 the Royal Society of Chemistry.

composition. Therefore, a combination of other spectroscopic methods, such as Raman spectroscopy, aids in the precise determination of the chemical composition of pigments.

The application of synchrotron X-ray radiation increases the sensitivity of XRF analysis. Access to high-energy synchrotron X-rays is limited due to the small number of radiation sources. Such high-energy XRF does, however, enable the obtaining of valuable information that would otherwise be impossible to acquire [5]. For example, Vincent van Gogh often reused the canvas of an abandoned painting and painted a new or modified composition on top [6]. Utilization of high-energy synchrotron X-ray fluorescence mapping allowed visualizing a woman's head hidden under Van Gogh's *Patch of Grass* work (Figure 2.8). X-ray fluorescence intensity maps allow accessing the distribution of specific elements (Hg and Sb) in the hidden paint layers. Such elemental mapping enabled the color reconstruction of the hidden face and linked it to Van Gogh's known portraits of women.

Another excellent example of using synchrotron X-rays for conducting XRF analysis is the investigation of the chemical composition of papyrus fragments found in the Villa dei Papiri at Herculaneum [7]. These papyri were carbonized during the eruption of Mount Vesuvius' in 79 AD and excavated between 1752 and 1754. Such XRF imaging has been performed in three different resolutions with 50 μm (low resolution), 10 μm (high resolution), and 1 μm (ultra-high

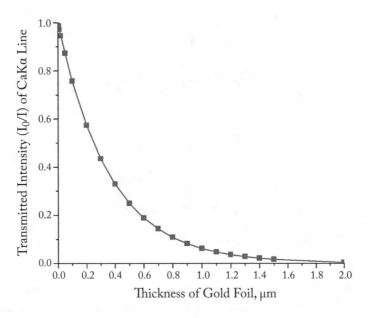

Figure 2.7: Dependence of I_0/I for CaKα line on gold foil thickness.

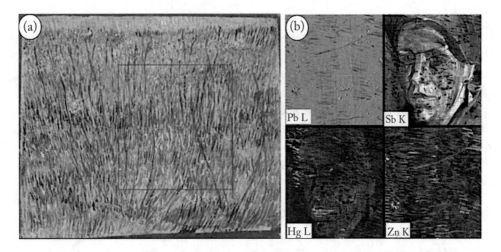

Figure 2.8: A painting of Vincent van Gogh (a) analyzed with synchrotron radiationbased XRF. Distribution of Pb L (preparation layer), Hg L (vermillion), Sb K (Naples yellow), and Zn K (zinc white) lines (b). Reprinted with permission from [6]. Copyright 2008 American Chemical Society.

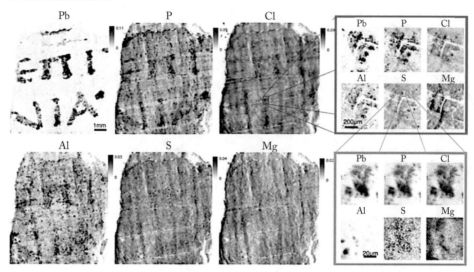

Figure 2.9: Multiscale XRF images of elemental distributions at low (left, 50-μm beam spot size), high (10-μm beam spot size, red frame) and ultra-high (1-μm beam spot size, green frame) resolutions. Reprinted with permission from [7] Copyright 2016 United States National Academy of Sciences.

resolution) beam spot sizes. Figure 2.9 shows that Pb is the main element common to all of the ink. At low-resolution imaging, P, Cl, and Al are also visible, while S is difficult to identify and Mg does not show any correlation with the ink. In the images obtained by ultra-high-resolution scanning, Pb, Cl, and P strongly correlated together, even though a high concentration of P can also be found in other places. The presence of lead in the ink, the authors of this research explained by several hypotheses. First, lead may have been used as a pigment (lead sulfide or lead white), or PbO (litharge) could have been used for a binder in the ink, as it speeds up oil drying. However, exact origin of Pb in the ink was hard to determine as the papyri were subjected to pyroclastic events and subsequent exhumation.

Nevertheless, XRF imaging helped to provide essential insights into the manufacturing and history of these papyri. The results of this investigation push the date back by several centuries for the use of metals in ink, back to Greco-Roman civilization. Prior to these measurements it has been a common belief that metals were not used in most ancient Greek and Latin manuscripts. The primary historic assumption was that carbon-based inks were used for writing until at least the 4th to 5th centuries AD, and that metallic (iron) gall inks were developed around 420 AD for writing on parchment, which is more adherent than papyri.

2.5   PARTICLE INDICTED X-RAY EMISSION (PIXE)

In PIXE, the atomic elements in cultural heritage samples are being excited through irradiation with a proton beam of typically more than 2 MeV energy from a small particle accelerator as discussed in Section 1.9. Higher energies help to excite the elements in deeper layers of the materials. The excited atoms decay by emission of their characteristic X-rays which are measured using traditional X-ray detector devices. A special application is the particle-induced γ-ray emission (PIGE) which relies on the excitation of the atomic nucleus and its subsequent γ-emission. PIGE is therefore sensitive to the isotopic composition of the material. Because of the much smaller cross section of these reactions, PIGE requires much higher beam intensities which can be destructive for the material. The method is therefore rarely applied in cultural heritage material analysis.

2.5.1   GENERAL CONSIDERATIONS

PIXE measurements can be procured in the air and do not require special sample preparation. This allows for characterizing different parts of objects. The automated analysis of a large number of samples, low detection limits, and excellent sensitivity for light elements (S, P, Cl, K, and Ca) make PIXE of great importance in archaeological sample characterization. To conduct an efficient analysis of multiple elements, two detectors can be used. In such cases, the high energy detector can measure heavy elements (from Fe to U), while a low energy detector allows measuring of light elements (from Na to Ni). These features enable the simultaneous determination of a considerable number of elements (10–20). X-ray spectra in the PIXE method have a low background, in comparison to the EDX method. Therefore, the detection limits are low (10^{-8}–10^{-10} g), making PIXE a perfect method for trace element analysis. Variation of the incident beam angle or beam energy allows for performing differential PIXE, to sample composition at different depths. Although the X-ray signal can come from different layers, the contribution from the surface is more abundant. Therefore, numerical methods should be used to differentiate contributions from shallow and deeper layers. The sampling depth depends on the composition of the material and can be reached 10–50 μm below the object's surface.

The most common PIXE equipment includes: an electrostatic accelerator with an ion source, with energies of up to 50 MeV; ion beamlines; in-beam diagnostic techniques, such as the Faraday cup; beam-focusing magnets; detectors with associated electronics for processing the signal and for data acquisition. In the microprobe analysis configuration, the sample is irradiated with a microbeam (1–50 μm). By scanning the beam on a part of the sample, high-resolution (limited by size of the beam spot) 2D or 3D elemental distributions can be determined. In such measurements, the signal from the detector should be recorded as a function of the beam spot position.

In many cases, cultural heritage materials cannot be analyzed in a high vacuum environment due to their large size, irregular shape, and/or the presence of volatile components. In such cases, an external ion beam extracted from the high-vacuum beamlines is used to analyze the objects. Several different types of windows made from thin (0.1–0.2 μm) metallic foils, polymers, or silicon nitride are used for such purpose. These windows do, however, reduce the beam energies by 50–250 keV. Beam spot size in such arrangements can be 10–100 μm. The use of x-y-z position-controlled sample holders and multiple X-ray detectors allow for generating elemental concentration maps. To reduce the absorption of produced X-ray by the samples from the air, and to decrease the effect of argon from the air, the region of interest can be flashed with helium gas. PIXE can often be coupled with other ion-beam analysis methods, such as Rutherford backscattering. PIXE has many similarities with XRF. Both methods analyze X-rays generated from the objects. The PIXE, however, uses protons instead of X-rays as probing radiation. In PIXE, the change in the energy of the proton allows for performing differential analysis of the elemental composition of both the shallow (surface) and the deep layers of an object.

2.5.2   EXAMPLES OF PIXE

PIXE can be used in different ways, in order to identify the elemental composition of an object. For example, interesting work has been conducted to identify the pigments used in Landscape, a painting of Italian modernist Mario Sironi (1885–1961) [8].

Figure 2.10a shows the painting with ten point-marks indicating the analyzed regions. Figure 2.10b shows two PIXE spectra taken from areas 1 and 8 with green and brown colors, respectively. The two characteristic elements in these areas are Zn and Ba, suggesting that ZnO and ZnS + $BaSO_4$ compounds were used as a white pigment for painting the preparation layer. Cr present in area 1 is presumably chromium oxide (Cr_2O_3), a synthetic green pigment used in the 19th century. The iron observed in both spectra can be related to yellow ochre, Prussian blue $Fe_4(Fe[CN]_6)_3$, and mars red (Fe_2O_3). The presence of Ca is associated with calcite ($CaCO_3$) or gypsum ($CaSO_4 \cdot 2\,H_2O$). Lighter sulfur confirms the use of gypsum, while Si and K evidence the existence of some silicate admixtures in the pigments. Figure 2.10c displays normalized PIXE results in so-called sun-ray plots. Such a simple representation allows for illustrating the relative elemental composition of the specific analyzed area.

The differential PIXE technique is a non-destructive approach for determining the layer composition and ordering in paint layers. In this case, PIXE spectra of the same region of interest are recorded at different beam energies, such as 1, 2, 3, 4, and 5 MeV. The increase in beam energy increases the sampling depth and allows for performing analysis of deeper layers and for constructing depth profiles of a particular layer. The paint-layer thicknesses and compositions deduced from differential PIXE are qualitative only. To achieve such analysis, the investigator assumes that the applied layers in the painting are parallel and that each layer has a homogenous composition in the analyzed volume.

An excellent example of conducting such differential PIXE investigation is an analysis of the altarpiece *Pala Albergotti*, painted by Giorgio Vasari in 1567 [9]. This is a painting on wood, composed of several panels in a wooden frame. Researchers conducting this investigation aimed to determine the composition and thickness of a dark alteration layer on the surface of one of the

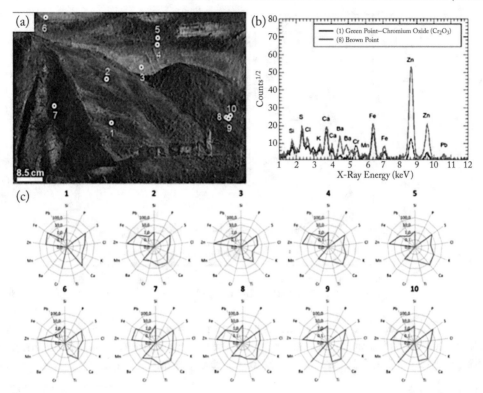

Figure 2.10: The oil painting Landscape by Mario Sironi with white point marks indicating the analyzed areas (a), PIXE spectra taken from 1 and 8 areas (b), and sunray plots of the PIXE results of all analyzed points (c). Reprinted with permission from [8]. Copyright 2014 Elsevier.

panels. Other panels were covered by a wooden frame for a long time (centuries) and maintain their brighter hue.

Analysis results for all of the "clean" green spots (non-altered areas with a brighter hue) suggested the use of a Cu-containing pigment. Dark green areas exhibit large amounts of Cu and smaller Pb and Sn content, while lighter green and yellow areas show more Pb and Sn. This fact allowed for concluding that a green Cu-based pigment was mixed with lead white, and lead-tin yellow was used to paint the light green and yellowish areas. Lighter elements (such as Al, Si, Ca, and Fe) were also detected. The researchers assumed that the detected Al, Si, and Fe were probably due to the deposit of atmospheric particulates, especially in the altered areas. Ca was assumed to be part of the preparation layer underlying the paint layers, as the lower intensity of the Ca X-ray was being more absorbed in the higher-density paint layer containing larger amounts of lead and tin. Figure 2.11b shows the Al, Si, Ca, and Fe concentration ratios as compared with Cu in the PIXE spectra taken at lower beam energies, pointing out their higher

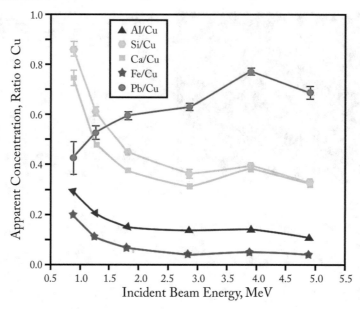

Figure 2.11: The ratios of elements to Cu as a function of the proton-beam energy, determined by differential PIXE analysis of an "altered" green area on the altarpiece *Pala Albergotti*, painted by Giorgio Vasari in 1567. Figure plotted from the data presented in [9].

concentration in the surface layer compared to the Cu-containing green layer. The presence over the green paint of such a surface layer is therefore verified through direct measurements by differential PIXE. Part of Pb is still detected from a layer below the one containing Cu.

The use of high-energy proton beams produced by particle accelerators can induce modifications in the most fragile components of the artworks. It was found that lead white and organic pigments, in addition to the binders, are the most sensitive components and can undergo irreversible damage upon high-energy ion irradiation [10]. It was suggested that harmless single-point PIXE measurements, with a 1 mm^2 diameter beam of 3 MeV protons, could be obtained using a beam with current below 0.1 nA for measurement duration of 100 s.

PIXE investigation of additional materials, such as alloys, has specific features compared to pigments and binders in the painting's layers. For example, differential PIXE was successfully used for probing the composition of silver alloys in Roman coins from Republic and Imperial periods (Figure 2.12a). Twenty (19 denarii and 1 antoninianus) coins, drawn from a private collection, minted between 136 BCE and 240 CE, were investigated by differential PIXE and XRF methods [11]. These measurements allowed the analysis of surface composition with different sampling depths. A proton beam with an energy of 3.4 MeV allowed probing the surface composition at ∼ 50 μm depth, while increasing the photon energy to 7 MeV increases the sampling depth to ∼ 140 μm. Increasing the beam energy further is challenging, as proton beam with higher energies would activate the copper through nuclear reactions and would, therefore, require extended post-analysis storage. At the same time, the induced activity of the samples initiates decays. In the latter case, PIXE can be complemented with the XRF analysis. In the case of silver alloys, XRF measurement allowed analysis of elemental surface composition with a sampling depth of $\sim 250\ \mu m$. Thus, the use of XRF with PIXE and different beam energies allowed for accurate depth profiling of elemental concentration (Figure 2.12b). The results show the average copper content for each coin as a function of time. The results indicated that copper content on the surfaces of D1–D4 coins is low, and increasing the sampling depth weakly influenced its Cu content. These results allow for concluding that the coins were minted using low copper content alloys. The measured Cu amounts in these four coins do not change as a function of sampling depth. This fact suggests that the surface modification of coins D1 through D4 is minimal. These results also indicate the stability of silver coinage content during the Roman Republic, despite the economic problems and the wars that led to the fall of the republic and the establishment of the Empire. Figure 2.12b silver coin demonstrates the debasements by Nero in 65 CE (coin D5) and by Trajanus (D8). Cu content relative to sampling depth indicated that debasement was accompanied by a significant surface modification, depleting copper content on the coin surface. The increase and fluctuation of Cu content at the end of the 2nd century CE and the first 40 years of the 3rd century CE provide evidence for the drastic devaluation policies established under Severian emperors (coins D13–D19). Significant differences in Cu, depending on sampling depth, for coins minted in 3rd century CE indicate extensive use of surface modification methods.

The stability of the Roman monetary system was influenced by coin availability and market demand for metals. The first debasement of denarii was conducted under Nero, and reflected in PIXE/XRF measurements, when the silver content was reduced to $\sim 90\%$. The Cu amount remained rather constant during the subsequent century, with some fluctuation (D8) that could be related to the debasements by Trajanus. The increase of copper amounts during the end of the 2nd century and the first 20 years of the 3rd century provide evidence for the drastic devaluation established under Septimus Severus, due to the increased costs of war expansion against the Germanic tribes and the Parthian empire. The political unrest during the last years of Septimius Severus' rule, and during the reign of his successors, is reflected in the fluctuation of the Cu content of the coins (D15 and D16) minted under Geta and Caracalla (198–217 CE). Caracalla developed a new taxation system, devalued the denarius, and introduced the antoninianus coin. Elagabalus abandoned the antoninianus but retained the debased standards of the denarius (coin D17). During the reign of Severus Alexander, some brief stabilization of the silver coins took place (D18). The further debasement of coins by Maximinus Thrax (235–238 CE) led to a further decrease in the silver content (D19). Eventually, the denarius was replaced by the antoninianus under Gordian III (238–244 CE), as the new standard silver currency (A20). The results of PIXE/XRF analysis reflect these historic facts and provides new knowledge on the modification methods that were used in Roman mints to produce coins with silver-enriched surfaces.

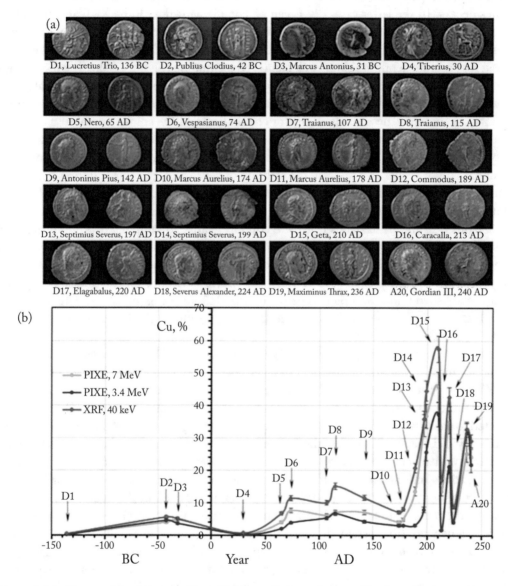

Figure 2.12: Photos (a) of Roman Republican to Imperial coins, D–denarii, A–antoninianus, and the measured copper content in the coins (b). Reprinted with permission from [11]. Copyright 2019 Elsevier.

The use of micrometer-sized proton beams in conjunction with position-sensitive x-y stages allows obtaining elemental distribution maps. For example, exciting research was conducted to investigate surface enrichment methods of several Roman republican coins [12]. PIXE mapping was performed with 3.0 MeV protons focused on a 3×3 μm^2 beam normal to the sample. The beam currents used were in the range of 0.1–0.2 nA. A silicon detector with an 80 mm^2 area and 145 eV resolution was mounted at 135° for detecting generated X-rays. Quantification and data treatment were performed with specialized software. Two types of silver coins, denarius and victoriatus, were cut to cross sections and were mapped by PIXE. Figure 2.13a shows Ag and Cu maps of a denarius with a highly homogeneous composition, 97.8% Ag and 1.4% Cu, and no signs of surface modification. On the contrary, the victoriatus coin that was analyzed presents an Ag-enriched surface layer. This coin contains at least three phases: a Cu-rich phase (63.7% Ag–35.6% Cu) and two silver-rich phases with 98.8% Ag–0.9% Cu and 79.4% Ag–19.7% Cu (Figure 2.13b). The thickness of the Ag-enriched surface layer is $\sim$ 150 μm. To produce such surface-modified metallic alloys, the Roman mints used several different approaches. One technique involved heating the silver-copper alloy in the air to oxidize copper at the surface layer. Then the alloy was soaked in vinegar (acetic acid) to dissolve the copper from the surface. This procedure created a pure but porous silver layer on the coin. Inverse segregation can be another method for coin surface modification. During the solidification of alloys, an Ag-rich layer can naturally be segregated, while the bulk alloy has a lower silver content. Other analytical methods should be used in conjunction with PIXE to determine the exact surface modification method used in such cases.

2.6 ELECTRON ENERGY-DISPERSIVE X-RAY SPECTROSCOPY (EDX)

The EDX method mainly utilizes an electron beam as an excitation source for the atoms in the sample to be investigated. The electron beam is generated in the scanning electron microscopes (SEM) or transmission electron microscopes (TEM).

2.6.1 GENERAL CONSIDERATIONS

This electron beam is focused on the object of interest and generated X-ray energies are measured by a detector that is cooled by liquid nitrogen or by Peltier cooling systems. Silicon-drift detectors are typically used in EDX analysis, as they have higher count rates and resolution. The detector is placed at an angle. The greater the angle between the detector and the sample, the higher the detection probability of the X-rays. The special resolution of the analysis is $\sim$ 1 μm. All elements from atomic number 4 (Be) to 92 (U) can be detected, and their relative concentrations can also be estimated by this method. EDS allows for determining the thickness of metallic coatings in multilayer materials and the analysis of composition of alloys. EDX analysis requires sample preparation, and only small volumes of the materials can be analyzed at one time. EDX can

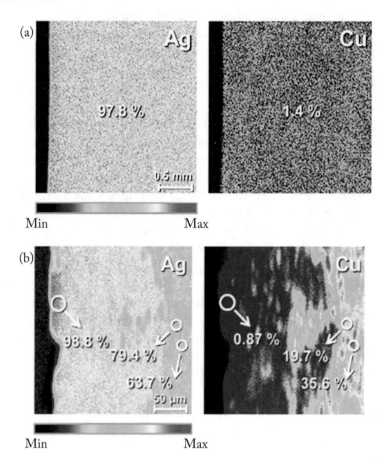

Figure 2.13: PIXE elemental Ag and Cu maps of the cut cross section of a denarius minted in 86 BC (a) and a victoriatus minted in 211 BC (b). Reprinted with permission from [12]. Copyright 2013 Elsevier.

be classified with distractive analysis methods, since a sample from an object is extracted and transferred into a high vacuum chamber. In some cases, samples should be subjected to special treatment (polishing, etching) before analysis. The benefit of such a destructive analysis is related to the accruing of additional morphological and structural information through simultaneous electron imaging methods. EDS also provides a high-resolution elemental mapping of samples.

2.6.2 EXAMPLES OF EDS

Chinese ancient lacquerware was produced by painting lacquer on different objects and occasionally the lacquer-coated objects were decorated with silver, gold, and pearls. The color of the

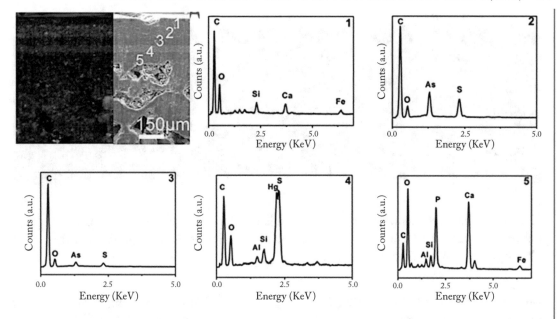

Figure 2.14: Optical and SEM images of cross-sectional samples (upper-left corner) and the EDS spectra of different layers. Reprinted with permission from [13]. Copyright 2016 Royal Society of Chemistry.

lacquer film was tuned by adding iron oxides (Fe_3O_4 and Fe_2O_3), lead oxide (Pb_3O_4), cinnabar (HgS), carbon, or other pigments. A lacquerware fragment sample dating back to the Song dynasty (AD 960–AD 1279) was analyzed by point-by-point EDX analysis [13]. Figure 2.14 shows the cross-sectional SEM image of the studied sample, indicating several layers of lacquer on top of a porous substrate. The surface layer (marked with #1) with black color contains C and O, as well as Si, Ca, and Fe traces. This composition suggests that the layer lacquer may consist of carbon-black pigment and oxide traces. The next point (#2) analysis in the yellow layer pointed out C, O, As, and S. The presence of As and S could be due to the existence of yellow-orange arsenic sulfide. A thin red layer (#4) consists of C, O, Hg, and S, suggesting the use of cinnabar. Between the red and the yellow layer (#3), a thin lacquer ash layer, containing carbon and oxygen and trace amounts of As and S, was detected. Beneath the red layer, the materials consisted of C, O, Ca, and P, indicating that it may be composed of bone and lacquer ash.

EDS mapping is a versatile tool for analyzing the elemental composition of various artifacts. For example, this method was used to investigate the corrosion of metallic and wooden parts of the Oseberg ship (Viking Ship Museum) [14]. This ship is considered one of the best artifacts to survive from the Viking Era. After excavation from a burial site in the early 20th century, a hot concentrated alum ($KAl(SO_4)_2 \cdot 12\,H_2O$) solution was used to conserve the deteriorated wood parts. This wooden ship also contains metallic parts such as nails and screws. Such

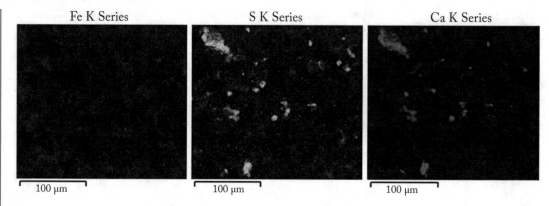

Figure 2.15: EDX elemental maps of Fe, S, and Ca in the corrosion product from a nail, showing S-rich areas mainly correlating with Ca-rich areas. Reprinted with permission from [14]. Copyright 2018 Macmillan Publishers.

alum treatment was prevalent during the early 20th century but it is damaging to the wood, as a sulfuric acid released over the years could degrade the wooden and metallic parts. EDX analysis was conducted to evaluate the effect of alum treatment on some metallic parts. This research showed that many metallic parts contained white corrosion powders, which were observed on the surfaces of the exposed rods. Samples of these powders consisted of potassium iron (III) sulfates, as confirmed by X-ray diffraction patterns. Such a composition of degradation product indicates the corrosive effect of alum treatment.

Meanwhile, corrosion products on some non-treated metallic objects were found to include iron carbonate ($FeCO_3$) and goethite ($FeO(OH)$), as detected by X-ray diffraction method. Such analysis showed no iron sulfate detected in the corrosion products. Sulfur and calcium were, however, observed by EDS mapping (Figure 2.15). These results allowed the authors to suggest some gypsum ($CaSO_4 \cdot 2\,H_2O$) presence in the corrosion layer, which is a common mineral that occurred naturally in Norway and was likely deposited during burial.

2.7   X-RAY DIFFRACTION

X-ray diffraction (XRD), though not a spectroscopic method, allows for the identification of crystalline materials. In crystalline solids, the constituent atoms (ions or molecules) are arranged in orderly three-dimensional crystal lattices. Non-orderly arrangements of constituents (absence of crystalline lattice) make solid materials amorphous. The same material can occur in both a crystalline and an amorphous form. For example, quartz is a crystalline mineral containing silicon and oxygen. Silicon oxide can also form amorphous structures such as glasses. Polymer resins, glasses, and organic binders in paintings are among the best examples of amorphous solids. Therefore, XRD analysis can be applied to identify crystalline objects.

2.7.1   GENERAL CONSIDERATIONS

Upon interaction of X-rays with ordered materials, a phenomenon known as interference can take place. Such interaction results in a pattern (diffraction pattern) that has lines with higher and lower intensities. The diffraction patterns obtained by XRD provide information about the internal structure of the material. In XRD analysis, a beam of X-rays is directed to a sample of interest, and the scattered intensity is measured as a function of direction, satisfying the Bragg's law:

$$n \cdot \lambda = 2 \cdot d \cdot \sin \theta, \tag{2.5}$$

where n is an integer $(1, 2, 3, \ldots)$, λ is the wavelength of the X-ray beam, θ is the scattering angle, and d is the interplanar distance between lattice planes of crystalline solids. The diffraction process occurs when an X-ray is scattered by the atoms (ions or molecules) of a crystalline solid (Figure 2.16a). The scattered X-ray waves can interfere constructively. In this case, two parallel waves remain in the same phase, and the difference between the path lengths of the two waves is equal to $n \cdot \lambda$. The path difference for two parallel X-ray waves undergoing constructive interference is equal to $2 \cdot d \cdot \sin \theta$. In XRD analysis, a diffraction pattern is obtained by measuring the intensity of scattered (constructively interfered) waves as a function of the θ angle.

X-ray diffractometers that allow recording of diffraction patterns consist of: an X-ray tube, a sample holder, and a detector. X-rays are generated in the tube, as described in Section 2.1. The spectra of the produced X-ray consist of several components ($K\alpha$ and $K\beta$). The wavelength of these characteristic X-rays depends on the target material (Cu, Fe, Mo, Cr) used in the tube. Copper ($K\alpha$ radiation $= 0.15418$ nm) is the most common target material for X-ray diffractometers. To produce monochromatic X-rays, filter foils or crystal monochromators are used. X-rays are directed onto the sample. During a typical pattern recording, the X-ray tube and detector slowly rotate simultaneously, while the sample is not moving (Figure 2.16b). When the geometry of the incoming X-rays satisfies the Bragg law, constructive interference occurs and a detector records a line. Crystalline materials usually have more than one crystallographic orientation. Each orientation could scatter X-ray at different θ angles. Therefore, a typical diffraction pattern consists of multiple rings (Figure 2.16c). Most crystalline solids have a unique set of diffraction lines. XRD pattern can be plotted as a profile, showing various lines with different intensities as a function of 2θ values (Figure 2.16d). For identification of specific material, recorded patterns are compared against known patterns stored in the database. XRD allows revealing specific materials compared to X-ray spectroscopic methods, which provide the elemental composition of the sample. XRD also allows the quantification and detection of minor phases, when the components in the mixtures are well crystallized.

X-ray diffractometry is a very popular method used for materials examined by science and chemistry. The use of commercial diffraction devices for art objects and artifacts is difficult, due to their large and irregular shape. Therefore, different types of dedicated devices have been developed that allow using the XRD method for analyzing materials in these objects. Portable

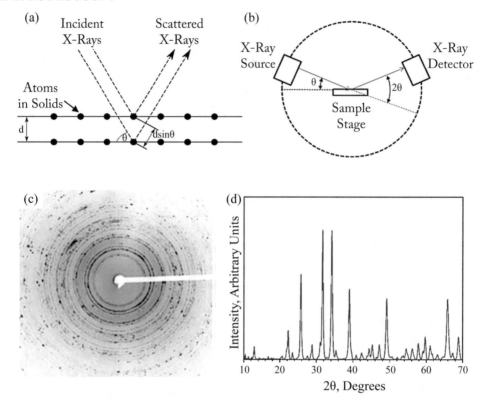

Figure 2.16: Scattering of two X-ray waves from the surface of crystalline solid (a), geometry of measuring the X-ray diffraction (b) an XRD pattern measured from a two-dimensional X-ray detector (c), and a profile showing lines with different intensities as a function of 2θ values (d).

devices (Figure 2.17) that combine XRD and XRF methods are used for artwork and artifact investigation [15].

Synchrotron radiation micro XRD is also used to probe the composition of paintings, pottery, and metallic artifacts. The use of synchrotron radiation-based XRD methods does, however, require preparation of the sample. Grazing incidence XRD is an alternative method for the investigation of the surface composition of objects. This modification also allows accurate depth profile investigation of layers with crystalline structure [16].

2.7.2   EXAMPLES OF XRD

An example is the use of the XRD method for identification of materials in medieval royal handwritten documents (royal solemn privilege) that date to 1459 and have been preserved in the Archive of the Royal Chancellery (Spain) [17]. The parchment document is written in Castilian

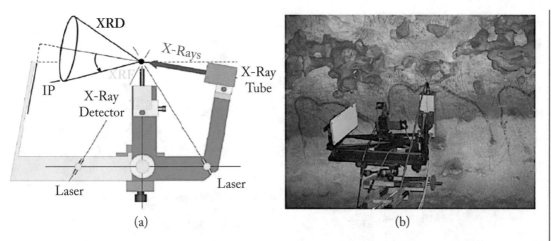

Figure 2.17: Schematic view (a) of a portable device XRD–XRF and its photograph (b) during measurement. Reprinted with permission from [15]. Copyright 2014 Elsevier.

with gothic letters on both sides, using a black-brownish ink with an illuminated initial capital letter, a small coat of arms (Figure 2.18).

This figure shows the XRD patterns from areas without (#1) and with text (#2). The analysis revealed that entire parchment sheets are covered with calcite, while gypsum also presents in the text areas. These results suggested that calcium hydroxide could have been used in the liming processes, to process the animal skin in parchment preparation. Over time, calcium hydroxide reacted with the atmospheric carbon dioxide to produce the calcite layer on the sheets. Gypsum was used to apply the preparation layer to the parchment surface before the application of the text, illumination, and gilding layers. XRD analysis suggests that calcite and gypsum are only present as crystalline phases. The text itself contains iron, as determined by XRF analysis, but this is not shown in diffraction lines, indicating the use of an amorphous ink. XRF analysis detected gold (#3) and silver (#4, #5, and #7) in decorative letters and different parts of the seal. Minium (Pb_3O_4), gold, silver, and calcite are the crystalline phases identified by XRD in the 4, 5, and 7 spots.

The use of micrometer-sized X-ray beams allows performing layer-by-layer XRD analysis. An excellent example of such micro-XRD analysis can be seen in the examination of a small cross section sample taken from an altarpiece named *The Visit of Saint Anthony to Saint Paul the Hermit* painted by Matthias Grünewald (Unterlinden Museum) [18]. Optical microscopy observation of this sample allowed identifying six distinct layers. Figure 2.19 shows XRD patterns of four materials, acquired from different areas of this sample. The preparation layer consists of calcite. The lead-tin-yellow (Pb_2SnO_4), hydrocerussite ($Pb_3(CO_3)_2(OH)_2$), and a small amount of cerussite ($PbCO_3$) are detected in the layers 2–5. XRF analysis showed a significant amount of copper in 2, 3, 4, and 5 layers. No diffraction patterns related to green-copper pigment were

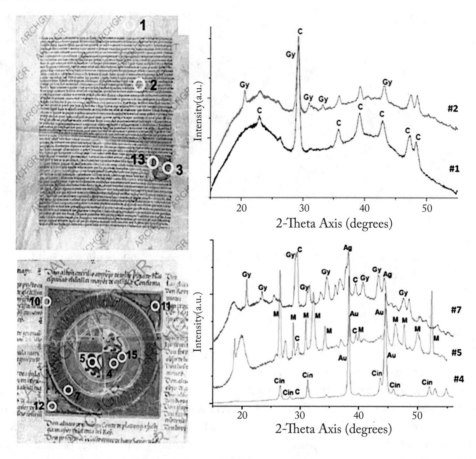

Figure 2.18: XRD patterns obtained on the specific areas on two parchments: C–calcite, Gy–Gypsum, Cin–cianabar (vermilion), M–minimum (Pb_3O_4), Au–gold, and Ag–silver. Reprinted with permission from [17]. Copyright 2014 Elsevier.

obtained, suggesting that the green-copper pigments are amorphous. The diffuse (broad) peak at low angles in the preparation layer XRD pattern corresponds to the scattering of X-rays by the resin used in sample preparation. Another broad peak obtained at the XRD pattern of area 4 can be related to amorphous green-copper pigments or organic binders. Patterns accrued from layers 2 and 3 show that the materials used in these layers are highly crystalline. This feature allowed for the use of Rietveld refinement analysis method, which obtained an approximate hydrocerussite/cerussite ratio of 3/1. Layer 6 did not exhibited XRD pattern and was assigned to amorphous organic varnish.

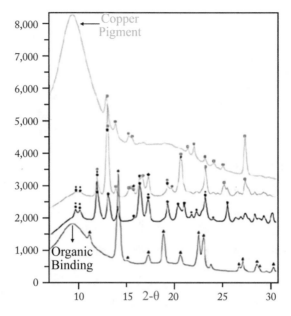

Figure 2.19: Layer-by-layer XRD analysis of a small cross sectional sample taken from an altar-piece named *The Visit of Saint Anthony to Saint Paul the Hermit* painted by Matthias Grünewald. Redrew based on a figure presented in [18].

2.8   RAMAN SPECTROSCOPY

In Raman spectroscopy, the sample is illuminated with a monochromatic laser beam measuring the inelastic scattering of light upon its interaction with vibrating molecules. The scattered light, with a frequency different from that of incident radiation, is used to plot a spectrum.

2.8.1   GENERAL CONSIDERATIONS

The Raman spectrum is an intensity wavelength shift diagram and can be recorded over a range of 10–4000 cm^{-1}. Raman spectrometers use light sources, filters, prisms (grating), interferometers, mirrors, and detectors (Figure 2.20).

The laser provides stable and intense beams of light with different wavelengths such as: 488 and 514.5 nm (argon ion laser), 530.9 and 647.1 nm (krypton ion laser), 632.8 nm (helium-neon), 785 and 830 nm infrared diode lasers, and 1064 nm (neodymium garnets). Argon and krypton ion lasers can cause significant fluorescence, making it challenging to record Raman peaks. Moreover, these lasers can initiate photodecomposition of the objects. Lasers with longer wavelengths reduce (or eliminate) fluorescence and can be used at higher powers without damaging the samples. Bandpass filters are used to isolate a single laser beam. Modern Raman spectrometers also use a combination of different filters and monochromators to separate weak

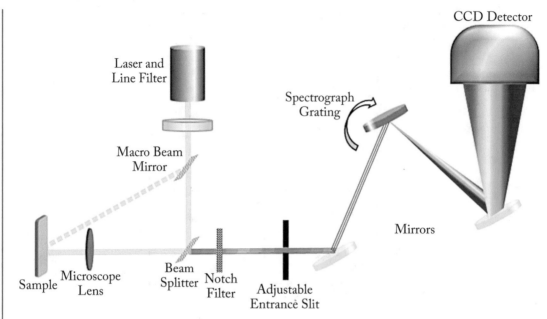

Figure 2.20: Schematics of a typical Raman spectrometer.

Raman peaks from Rayleigh scattered light. Spectrometers use arrays of sensitive CCDs that convert the optical signal into charge, integrating and transferring it to reading devices. Multi-channel CCD detectors are used with laser wavelengths of less than 1000 nm. In spectrometers with laser wavelengths of greater than 1000 nm, Ge or InGaAs detectors are used.

Combining Raman spectrophotometers with an optical microscope (Raman microscopy) provides for localized spectroscopic and morphological examinations. The optical lens helps to focus the laser beam onto the sample, and digital cameras enable photographing the region of interest. This method also allows mapping measurements and imaging of specific Raman peaks. Recently many different types of portable, hand-held, or standoff Raman spectrometry systems have been developed for archaeological and museum sample investigations. The use of short wavelength laser sources and optical fiber probes allows for remote analysis of samples. The use of surface enhanced Raman spectroscopy (SERS) significantly improves the sensitivity of weak Raman scattering. In this technique, the objects are covered with thin colloidal metallic films (silver or gold) to enhance the intensity of weak Raman signals. This approach also allows quenching the fluorescence.

In Raman spectroscopic investigations, the identification of materials in the unknown sample is mostly based on comparison of the recorded Raman spectrum with a library of reference materials. During the identification of materials, the comparison of the measured spectrum should be performed with the correct reference spectrum, which has been recorded with the same laser wavelength. Several research articles [19]–[21] provide spectral libraries of the Raman spectra for important inorganic pigments, organic colorants, and materials in archaeological artifacts. Quite a few online open-access extensive libraries, like http://rruff.info, www.minerals.gps.caltech.edu, and www.ens-lyon.fr/LST/Raman/index also provide reference Raman spectra for minerals.

2.8.2   EXAMPLES OF RAMAN SPECTROSCOPY

Raman spectroscopy is a prevalent method for the identification of materials used in the preparation of archaeological ceramics and pottery. It can be used to investigate the object's body, decorations, paintings, and glazes. The components in the body of ancient pottery originate from the raw material (clays) and additive. Identification of these primary minerals allows for revealing the main ingredients used to prepare the objects. The phases formed during the firing explain the technology and conditions of processing. Information on the presence of secondary compounds, formed during the use of objects and/or their burial, provides clues regarding the life of the object as well as the conditions during the burial.

An example of using Raman spectroscopy to determine the mineralogical composition of two pottery samples (named A and B) can be found in the analysis of Bronze Age Cypriot ceramics from the collection of the Ringling Museum of Art (Sarasota, Florida, USA) [22]. Quartz, a modification of silicon oxide with the SiO_4 tetrahedral units linked to each other, was found in these objects. Raman spectrum of a quartz crystal contain an intense band representing the symmetric bending vibration (Si–O–Si) at 463 cm^{-1}, with some other weak bands at 126, 200, and 262 cm^{-1} (lattice vibration modes) and bands at ~ 357 and ~ 400 cm^{-1} corresponding to the asymmetric bending modes of the SiO_4 group. The spectrum also contains two weak bands due to the Si–O–Si bending mode (~ 802 cm^{-1}) and an asymmetric stretching vibration of the SiO_4 (1085 cm^{-1}).

Both samples investigated exhibit a significant amount of anatase, which is the low temperature crystal form of titanium oxide (TiO_2). This compound shows an intense Raman band at ~ 145 cm^{-1} in addition to four weaker peaks at 196, ~ 395, 515, and ~ 640 cm^{-1} (Figure 2.21). It is highly likely that anatase was part of the clay used for making the pottery vessels. The presence of this crystal modification suggests that the firing temperature was below the anatase-rutile transition point ($850 \pm 100°C$). Rutile is the high temperature modification of TiO_2 and can be formed by heat treatment of anatase. Traces of rutile micro-crystals were also detected in the samples. This modification of TiO_2 shows two bands centered at ~ 447 and ~ 610 cm^{-1} (Figure 2.21). The presence of a significant amount of anatase and only traces of rutile allows concluding that the rutile is part of the clay and not a result of the firing of the anatase-rich clay. Calcite was also identified in samples, based on the presence of four bands corresponding to the Raman-active modes of calcite. Calcite can be used as an additive in clay. Its presence allows confirming that the pottery was heat-treated below 950°C, which is the decomposition temperature of calcite.

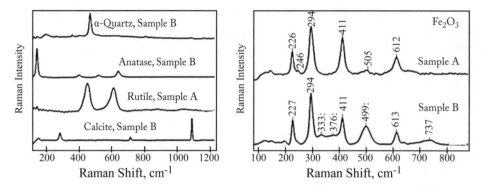

Figure 2.21: Raman spectra obtained from different regions of pottery samples A and B. Redrew based on a figure presented in [22].

The red pigment on the pottery is mainly identified to be hematite (α-Fe_2O_3). According to the space-group theory, the hematite has seven Raman-active vibrational modes. Six of those vibrations were found in the Raman spectrum (Figure 2.21). The broader band centered at $\sim$ 400 cm^{-1} in the spectrum for Sample B and small bands (marked with asterisks) at 350 and 737 cm^{-1}, were attributed to maghemite, γ-Fe_2O_3, a different modification of iron (III) oxide. These two modifications have different structures and have, therefore, distinguished Raman spectra. The presence of Fe_2O_3 in the pottery samples suggests that heat treatment was conducted in an oxidizing atmosphere.

The position shifts and broadening of hematite peaks can be used to distinguish natural hematite from that obtained by the heating of goethite (α-FeOOH). These features also permit estimating the presence of impurities (such as Al) in the hematite, evaluating the degree of crystallinity, and relating it with the raw material's provenance and firing conditions. The identification of red and black color iron-oxides is a suitable approach for differentiating the heat treatment in the reducing or oxidizing atmosphere. However, there is debate about the origin of a band at $\sim$ 670 cm^{-1}. In some cases, this band is attributed to the presence of magnetite (Fe_3O_4). The identification of red hematite indicates an oxidizing environment, while the black Fe_3O_4 is obtained under reducing conditions. Different works also suggested that the band at $\sim$ 670 cm^{-1} could be present in amorphous hematite. Therefore, it is challenging to identify Fe_3O_4 based on this peak. On the other hand, a single band positioned at $\sim$ 660 cm^{-1} can be attributed to magnetite. However, the position of this peak may be shifted depending on impurities.

The identification of pigments on artworks and artifacts by Raman spectroscopy allows revealing their preparation methods and helps in restoration, conservation, dating, and attribution efforts. The combination of Raman spectroscopy with an optical microscope allows performing non-destructive analysis with high spatial ($\sim$ 1 μm) and spectral ($\sim$ 1 cm^{-1}) resolution. An ex-

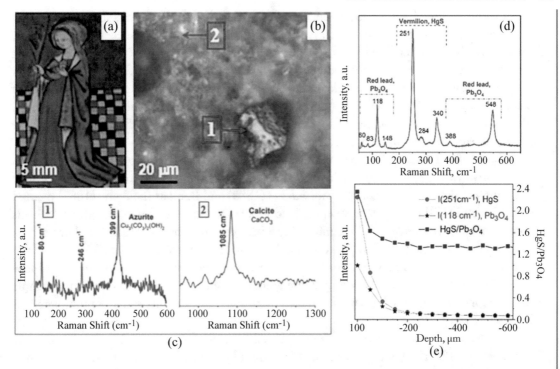

Figure 2.22: Photograph of manuscript page (a), magnified optical microscopy image in a blue area (b), and Raman spectra of blue and white pigments (c), as well as Raman spectra for red areas (d), and depth profile results (e), for HgS and Pb_3O_4 in the red area. Reprinted with permission from [4]. Copyright 2016 the Royal Society of Chemistry.

ample of combined use of the Raman spectrometer and the optical microscope can be found in an investigation of an illustration (Figure 2.22a) of a 15th-century Breton manuscript [4]. The optical microscopy image (Figure 2.22a) of blue areas shows that there are two pigments with blue and white colors. The Raman spectrum (Figure 2.22c) of the blue particle shows characteristic bands for azurite ($Cu_3(CO_3)_2(OH)$) mineral. The white area exhibits a peak centered at ~ 1085 cm^{-1} associated with a vibration mode of calcite (Figure 2.22c). Investigation of the different blue regions suggests that the artists changed the ratio of azurite and calcite to achieve different shades of blue color.

Raman spectrum of the red area in Figure 2.22 suggests a mix of red lead (Pb_3O_4) and vermilion (HgS). This spectrum shows that the intensities of the HgS peaks are higher than those of Pb_3O_4. XRF analysis of this area indicates that the concentration of lead is nearly five times greater than mercury (Figure 2.22d). Raman spectroscopy provides a depth-profiling feature, which is helpful in such cases, to understand conflicting results [23]. Figure 2.22e presents the depth profiles of the band at 251 cm^{-1} (HgS) and the band at 118 cm^{-1} (Pb_3O_4), through the layer of paint, and the HgS/Pb_3O_4 intensity ratio for these bands. This ratio allows concluding that the surface layer of the paint (~ 150 μm) contains vermilion and its concentration gradually decreases within that layer. The HgS/Pb_3O_4 ratio in the second layer ($-150–600$ μm) is smaller. These results may suggest that the red area was painted with Pb_3O_4, and then the HgS layer was applied. It is possible that the artist used such bilayer paint to achieve a distinctive shade of red color. The application of multiple paint layers could also be related to the price of pigments. HgS was a more expensive pigment than red lead and the artist probably first painted a thicker layer of red lead followed by an application of a thin HgS layer.

The localized nature of Raman microscopy also allows investigating small samples taken from the artworks. For example, such investigation was performed on the cross sections of yellow paint samples taken from *View of Arles with Irises* (1888) and Bank of the Seine (1887) painted by van Gogh [24]. The samples were embedded in a resin and ground with Micromesh sheets before inspection (Figure 2.23). The microscope images show that both samples have a dark varnish layer applied to the yellow paint layers. A sample taken from the *View of Arles with Irises* is characterized by a complex composition with a main yellow paint layer, on top of a blue-white-yellow alteration layer containing some greenish grains. The cross section of the sample of *Bank of the Seine* consists of a top brown layer with some black grains and an inhomogeneous yellow-white layer.

The Raman spectrum acquired from the yellow area of *View of Arles with Irises* shows chrome yellow ($PbCrO_4$) with symmetric stretching (841^{-1}) and ~ 406, 380, and 361 cm^{-1} bending vibrations for CrO_4^{2-} group. The complex yellow-white-blue areas revealed the presence of $PbCrO_4$, lead white, and Prussian blue ($Fe_4[Fe(CN)_6]_3$) and greenish grains of Paris green ($Cu(C_2H_3O_2)_2 \cdot 3\,Cu(AsO_2)_2$). The yellow-white areas for *Bank of the Seine* are composed of a mixture of lead white (a band at 1052 cm^{-1} of CO_3^{2-} symmetric stretching) and chrome yellow.

The identification of organic and biomolecules when using regular Raman spectroscopy is challenging. Surface-enhanced Raman spectroscopy (SERS) allows significant ($\sim 10^{10}$) enhancement of Raman scattering process, using colloidal metal surfaces. These surfaces make it possible to detect even single molecules. In this method, small samples extracted from the artwork are covered by a thin layer of metal, such as silver. The lithography Ag nanoparticles, silver nano-island films, and colloidal Ag nanoparticles prepared by solution reduction can be used as SERS support.

An example [25] of the application of SERS for the analysis of artwork is the investigation of samples taken from an Egyptian polychrome leather fragment (Figure 2.24a) dated to the Middle Kingdom (2124–1981 BC). The red translucent paint on the object contains an organic dye. SERS analysis of a red glaze revealed the spectrum shown in Figure 2.24c, corresponding with madder lake extract. This dye is one of the most stable natural pigments and was extracted from the madder plant's root. This pigment was in everyday use by the ancient Egyptians for coloring textiles. In the Middle Ages, the madder plant was cultivated in Europe to extract

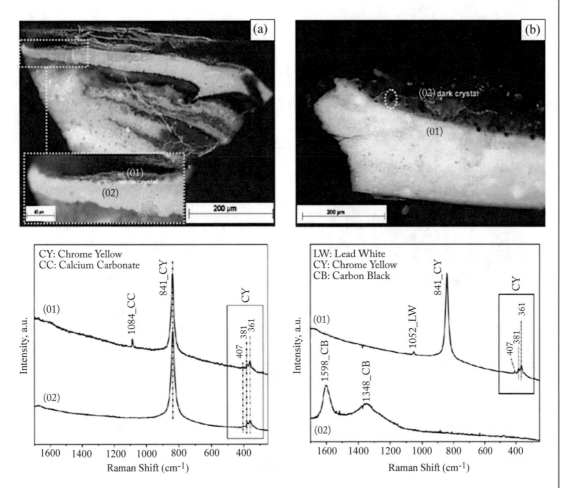

Figure 2.23: Optical microscopy image (top) and Raman spectra (bottom) for van Gogh painting samples from *View of Arles with Irises* (panel a) and *Bank of the Seine* (panel b). Reprinted with permission from [24]. Copyright 2011 American Chemical Society.

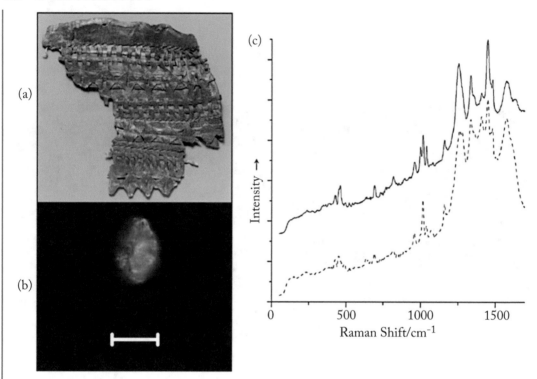

Figure 2.24: SERS analysis of Egyptian painted leather fragment. Photograph of the object (a), and (b) polarized reflected light image of sample removed from red painted area (scale bar is 20 μm). SERS spectrum of sample (c). Solid line: spectrum of sample; dashed line: spectrum of a pink pigment from madder lake reference. Reprinted with permission from [25]. Copyright 2009 United States National Academy of Sciences.

dye for textile coloring. However, it was not used in medieval or Renaissance paintings. The discovery of madder organic dye in the 4,000 years old object is the earliest evidence for the use of sophisticated chemical methods to extract a natural colorant from a plant.

2.9   FOURIER-TRANSFORM INFRARED (FTIR) SPECTROSCOPY

In FTIR spectroscopy the sample is exposed to infrared light, which transfers much less energy and therefore won't cause atomic excitation processes. It will however cause molecular vibrations (heat transfer), which reflect on the chemical structure of the materials in the sample. FTIR spectroscopy therefore emerged as a widely used technique for the investigation of artworks and historical artifacts.

2.9.1   GENERAL CONSIDERATIONS

Upon exposure to IR radiation, the functional groups in molecules absorb light, causing changes in the dipole moment. This absorption results in vibrational energy level transfer from ground state to excited state. The number of light absorption peaks is determined by the degree of vibrational freedom of the molecules, while the intensity of the peaks depends on change in the dipole moment and the transition of energy levels.

The spectrometer consists of an IR radiation source, an interferometer, a sample holder, a detector, an amplifier, an analog-to-digital converter, and a computer. In conventional FTIR analysis, the solid-state materials (powders) are dispersed in infrared inactive powder (potassium bromide), prepared as pellets and placed between infrared transparent windows. During the measurements, the light passes through the sample. The IR radiation passes the interferometer and the sample then reaches the detector.

The interferometer is a critical component in FTIR spectrometers; it splits the light beam into two different paths (Figure 2.25a). The interferometer contains a beam splitter and two (one stationary and one movable) perpendicularly aligned mirrors. The beam splitter transmits $\sim$ 50% and reflects about 50% of the radiation emitted from the source.

The transmitted light strikes the stationary mirror, while the reflected light strikes the movable mirror. The detector measures the intensity differences for the two beams as a function of the optical path difference (OPD or δ). The maximum intensity signal can be obtained when $\delta = n\lambda (n = 0, 1, 2, \ldots)$ where the λ is wavelength. When is δ equals multiples of half-wavelength, destructive interference occurs, resulting in minimum intensity signal $\delta = (n + 1/2)\lambda$. Due to the movement of the moving mirror, the intensity of the signal oscillates between maximum and minimum values of OPD (in centimeters), resulting in a cosine wave known as an interferogram (Figure 2.25b). When the distances of the two beams are the same, we describe the situation as zero path difference (ZPD). The use of a radiation source that provides light with different wavelengths (broadband source) results in a single intense peak at ZPD on the interferogram (Figure 2.25b). The interferogram is a function of time. To obtain a spectrum from interferogram, a mathematical method to transform a function into another function—known as Fourier transform—is needed. The time domain of the interferogram should be Fourier transformed in order to obtain a frequency domain. During this mathematical operation, inversion of the OPD is performed. Since the unit of OPD is the centimeter, after inversion, the new function will have a unit of inverse centimeters, or cm^{-1}, also known as wavenumbers. Fourier transformed interferogram is an IR spectrum of broadband light and represents a plot of signal vs. wavenumber (Figure 2.25c).

During typical IR measurements, the IR spectrum of broadband light is recorded first as background; then, the sample is placed between the beam splitter and detector. The interferogram with the sample is recorded and the background spectrum results in the IR spectrum of the studied sample are subtracted. The IR spectrum can be presented in absorption (A) or

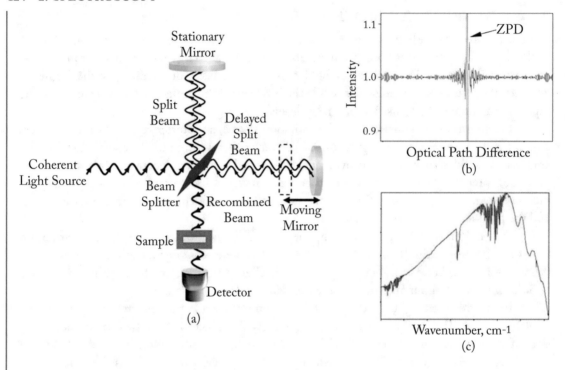

Figure 2.25: Schematics of a FTIR spectrometer (a), a typical broadband light source interferogram (b), and its IR spectrum (c).

transmittance (T) intensity (in %) plotted against the wavenumber:

$$A = \frac{1}{T} = -T = -\frac{I}{I_0},$$

where the ratio of power transmitted by the sample (I) relative to the power of incident light on the sample (I_0) results in a quantity of T, while the absorbance (A) is the logarithm of the reciprocal of T; samples that are IR opaque can be analyzed in the reflectance mode. However, if the samples' surface is not reflecting, the interpretation of acquired spectra can be challenging because of the presence of refraction, scattering, or diffuse reflection. To overcome this problem, the attenuated total reflection (ATR) method can be used, which uses a crystal with a higher refractive index than the analyzed sample. In ATR-FTIR spectroscopy, solid samples can be placed onto an internal reflection element, also called an ATR crystal (Figure 2.26). The IR beam is directed onto the crystal at an angle that is greater than the critical angle. This allows infrared light to undergo internal reflection. At each point of reflection, a short-lived (evanescent) wave is produced that can be absorbed by a sample. Such an arrangement allows investigating the absorption of the infrared light by the sample in direct contact with the crystal. The absorbance

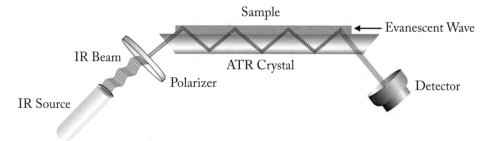

Figure 2.26: Schematics of ATR-FTIR system.

of IR light is proportional to the effective path length, which is equal to the total number of reflections of the IR beam times the penetration depth in the sample.

2.9.2   EXAMPLES FTIR SPECTROSCOPY

FTIR spectroscopy is a suitable way to explore the degradation of materials in artworks. For example, deterioration of external wall paintings is a complex process due to the multiple environmental factors (extreme temperatures, UV radiation, and humidity), heterogeneous nature, and original condition of the paintings. The presence of soluble salts, the interaction of pigments with binders, microbiological activity, and other factors complicate the degradation pathways of the paintings. Revealing the degradation products allows understanding of the degradation mechanism and identification of the original materials, and helps in restoration efforts. An example of such an attempt can be found in the investigation of small samples taken from exterior wall paintings of Agios Sozomenos Church in Galata (Cyprus) [26]. These wall paintings (Figure 2.27a) have high significance to the history of Cypriot mural decoration during the post-Byzantine period. Analysis of cross-sectional samples, using FTIR spectroscopy in reflectance mode, suggested the significant presence of calcium oxalate minerals ($CaC_2O_4 \cdot 2H_2O$ and $CaC_2O_4 \cdot H_2O$).

Analysis of the blue-green layer marked within the yellow box in the cross-sectional optical image (Figure 2.27b) suggests the presence of both copper and calcium oxalates (Figure 2.27c), evidenced by the strong $C = O$ stretching vibrational bands at 1620–1660 cm^{-1}. The FTIR spectrum (Figure 2.27c) of a reference calcium-oxalate compound confirms the presence of this compound. The identification of moolooite (copper oxalate) is based on the additional peaks centered at 1360 cm^{-1} and 1320 cm^{-1}. Blue and green copper carbonate pigments derived from azurite and malachite minerals have been used extensively in wall paintings. Cyprus, named for the copper element, is rich in copper mineral deposits such as copper carbonates, and the use of such carbonate-based pigments is expected in the paintings. The identification of oxalates in the paint sample could imply that original pigments degraded over time to produce moolooite and calcium oxalates. Copper oxalate has never been used as a pigment. A possible pathway for

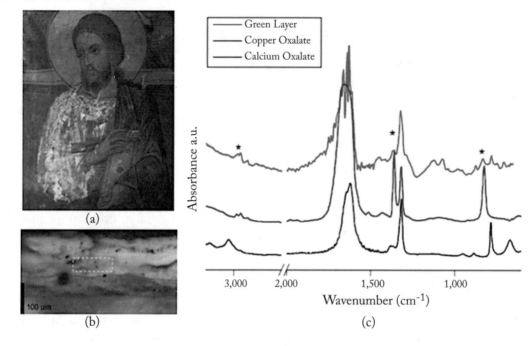

Figure 2.27: Painting of an Apostle from the early 16th-century exterior wall painting of Agios Sozomenos Church (a), a cross-sectional sample from the flaking green paint (b), and FTIR spectra of the blue-green layer and reference materials (c). *marks show the peaks assigned to copper oxalate. Reprinted with permission from [26]. Copyright 2008 Elsevier.

the formation of oxalates could be slow degradation of binders, such as proteins or other organic compounds, triggered by lichens to form oxalic acid. A lichen-induced oxalic acid could react with the copper carbonate pigments and calcium sulfate minerals used in wall preparation.

The slow reactions of pigments with organic binders in paint layers can cause alterations, such as significant loss of mechanical integrity and color changes. An example of a frequent, pigment-binder interaction is the saponification process. This chemical reaction produces water-insoluble compounds with metals and free carboxylic acids originating from oil binders. Lead- or zinc-containing pigments, for example, react with carboxylic acids producing metal soaps or salts in the form of aggregates, crusts, and blisters, among other types of new formations. Well-demonstrated examples of such interactions are found with oil paintings containing lead white, red lead, lead-tin yellow, and zinc white (ZnO). The hydrolysis of triacylglycerides in oil binders results in the formation of free carboxylic acids (fatty acid). These acids react with the metal-containing pigment particles to form metal soaps.

Another pigment-binder interaction is the reaction between egg yolk binder and lead-tin yellow pigment found in a 15th-century Gothic panel painting [27]. To increase the understanding of such processes, a model investigation was conducted to study the reactions between lead-tin yellow I and three different binders: egg yolk (EY), an emulsion of egg yolk with poppy seed oil (PO), and a mixture of lead-tin yellow I with poppy seed oil. These binders were mixed with Pb_2SnO_4 in plastic vials and reacted for up to 6 months. FTIR spectroscopy was used to investigate the products after aging for 1, 3, and 6 months.

In the lead-tin yellow-egg yolk suspension (Figure 2.28a), gradual disappearance of triacylglycerides bands (arrows) and the simultaneous formation of bands (highlighted area) for a new lead carboxylates phase indicates saponification reaction. Such transformation is more distinct in the Pb_2SnO_4 + EY + PO mixture (Figure 2.28b), while FTIR spectra for Pb_2SnO_4 + PO shows little change (Figure 2.28c). These results suggest that the combination of egg yolk and oil significantly accelerates the saponification reaction.

2.10   ATOMIC ABSORPTION, EMISSION, AND MASS SPECTROMETRY METHODS

There are several different micro-destructive atomic absorption and mass spectrometry analytic methods used in materials characterization for cultural heritage and historical artifacts. These methods include Atomic Absorption Spectroscopy (AAS), Laser-Induced Breakdown Spectroscopy (LIBS), and Inductively Coupled Plasma Mass Spectrometry (ICP-MS). These methods provide an analysis of the elemental composition of materials.

2.10.1   GENERAL CONSIDERATIONS

AAS allows for the quantitative measurement of chemical elements using the atomic absorption of light by free metallic ions in the gas state. Vaporization of a solution or solid sample allows simultaneous determination of multiple elements. During atomic absorption, ground state atoms absorb energy (light) and are elevated to an excited state. The amount of energy absorbed by atoms can be measured and used to determine concentration in the sample. The relationship between the amount of energy absorbed and the concentration of samples can be determined when compared with similar measurements performed with known standards.

The essential components of this spectrometer are the light source (multi-element hollow cathode lamps), a monochromator, a solid-state detector to measure the light, electronics, and a data display system. To atomize the sample, air/acetylene or nitrogen-oxide/acetylene flame is commonly used. This analytical configuration is relatively inefficient and only a small fraction of the sample can be analyzed. Atomization of the entire volume of the sample and interaction with light for an extended period significantly enhances the sensitivity of the technique. In inductively coupled plasma (ICP) technique, the interaction of a radio frequency field creates plasma with temperatures as high as 10,000°C. This technique allows complete atomization of the elements in a sample.

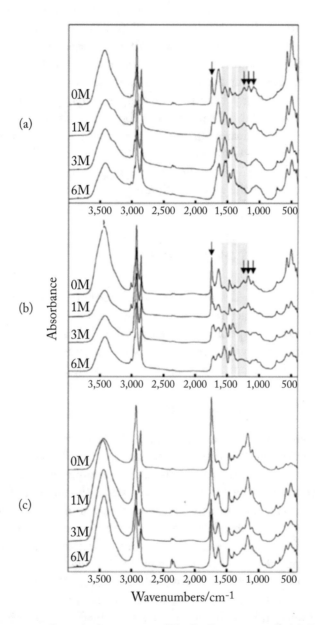

Figure 2.28: FTIR spectra of material mixtures of Pb_2SnO_4 + egg yolk (a), Pb_2SnO_4 + egg yolk + poppy seed oil (b), and Pb_2SnO_4 + poppy seed oil (c) collected after 0, 1, 3, and 6 months (M) of reaction. Triacylglyceride band (arrows) and lead carboxylates (highlights). Reprinted with permission from [27]. Copyright 2019 Elsevier.

Optical emission spectroscopy with ICP is used to efficiently analyze the light emitted from the elements in a sample. LIBS is an atomic emission spectroscopy, which uses a localized laser pulse to form a plasma to atomize the samples. The use of a high-power laser allows rapid analysis of samples. Upon contact of the laser with the object, the ablation of surface atoms creates a high-temperature (up to 10000°C) and short-lived plasma. At high temperatures, electrons in outer orbitals of atoms are excited. Upon cooling, the excited electrons move to lower orbitals, emitting photons. The detection limits in this method are a function of several different factors such as plasma excitation temperature, the light collection window, and the line strength of the transitions. Different types of lasers can be used in this method. The Nd:YAG source with 1064 nm wavelength is the most common laser used. A short ($\sim$ 15 ns), single laser pulse or multiple pulses with repetition rates up to > 20 Hz are used to ablate and excite atoms on the surface of a sample. The photons from the created plasma are collected using a lens and transmitted to the spectrometer, then analyzed by a CCD detector. LIBS is a rapid method and requires no sample preparation, but it is a micro-destructive method. Different handheld devices are available with high sensitivity to the light elements.

ICP-MS is a mass spectrometry method that uses plasma to ionize the sample. The argon plasma generates singly charged ions, from the elemental species in the sample, that are directed and separated by a quadrupole mass spectrometer. The detector analyzes separated ions with different mass-to-charge ratios. The analysis efficiency of this method is much higher than classical AAS, in addition to having low detection limits. ICP-MS also allows detecting different isotopes of the same element. The duration of analysis, precision, and sensitivity of the ICP-MS method is higher than for absorption spectroscopy. There are several strict requirements for ICP-MS. The total dissolved solids content of the analyzed sample should be very low to ensure maximum measurement stability. The instrumentation parts need to be cleaned periodically to maintain acceptable performance. ICP-MS can simultaneously determine 70 elements. ICP-MS requires sampling from the objects, desolation, and digestion of samples into solvents. Laser ablation, coupled with ICP-MS, is also applied to increase the efficiency of analysis. For quantity analysis, a well-known standard of the analyzed element should be used. The examples below will be limited to laser ablation ICP-MS as this method requires minimal sample preparation for cultural heritage objects.

2.10.2   EXAMPLES ABSORPTION, EMISSION, AND MASS SPECTROMETRY METHODS

Glass melting and production date back to the Late Bronze Age (16th century), which originated as a region that is now Syria and Northern Iraq. Historical glass preparation involved three principal raw materials: quartz sand, plant ash as an alkali metal source, and colorants. Calcium oxide prepared from annealing of lime was also added. This type of glass is called soda-lime glass. The glass was considered a high-status material and was rare until the end of the 17th century, making historical glass an infrequently analyzed object.

Sample preparation for laser ablation ICP-MS involves embedding small glass fragments in epoxy resin, followed by polishing the surface in order to remove an up to millimetric-thick contaminated surface layer. Large glass fragments can be directly analyzed. In this case, a pre-analysis ablation scan may be used for exposing fresh surfaces. For quantification and calibration, NIST glass standards (NIST 610 and 612), or the Corning glasses A, B, C, and D can be used, which mimic the composition of Egyptian and Mesopotamian, Roman, and Medieval glasses [28].

An exciting example of such analysis can be the laser ablation ICP-MS examination of medieval Islamic glass from Samarra, the capital of the Abbasid Caliphate (836–892AD) [29]. Samarra was palace-city and home to glass production and trade. The palace and city are known for their abundance of decorative glass and represent an essential source of archaeological information on early Islamic art and architecture. The authors of this study examined more than 250 glass artifacts (Figure 2.29a–e) from several different collections in European museums. Trace elements in the glass, identified using mass-spectrometric analysis, offered clues to the geographic origin of the raw materials used in the making of the different types of glass artifacts. Al_2O_3 vs. MgO/CaO ratios were determined (Figure 2.29f) and compared with published data of glasses in earlier studies from Ramla, Tyre, Nishapur, and Veh Ardasir (see citations in the [29]), indicating separation lines between early Islamic glasses from the eastern Mediterranean with low MgO to CaO ratios, Samarra glass with high MgO/CaO and low Al_2O_3 levels, and Mesopotamian glasses with higher Al_2O_3 concentrations typical of central Asian production. The results suggest that a part of Samarra's glass was imported from other areas, such as the Levant (Eastern Mediterranean) and Egypt. However, the majority of the artifacts were similar in composition, suggesting that the glass was being produced locally.

Another example of tracing the origin of materials can be an investigation of obsidians by ICP-MS. Obsidian is naturally formed volcanic glass. Obsidian glasses have colors ranging from green to gray and black. High hardness and sharp edges when fractured allowed humans to prepare tools with obsidian during the Stone Age, before the development of metallurgy. The characterization of obsidians from different archaeological areas allows for establishing the source of the excavated tools. Laser ablation ICP-MS allows isotopic ratio determination of obsidian samples and historical obsidian artifacts. This method provides an isotopic ratio precision of 0.002% relative standard deviation under optimum conditions. Also, ICP-MS offers high samples throughout and permits isotopic analysis of elements with high ionization potential. This feature allows precise tracing of origins for obsidian samples. For example, the primary known obsidian sources of the different Mediterranean locations (Monti Arci of Sardinia, Lipari, Palmarola, Pantelleria, Yali, Melos, Antiparos, Carpathian) are well known (Figure 2.30a). Using a log ($^{88}Sr/^{93}Nb$) isotope ratio analysis enables the classification of samples of known geological origins and facilitates attribution of origin for unknown samples [30].

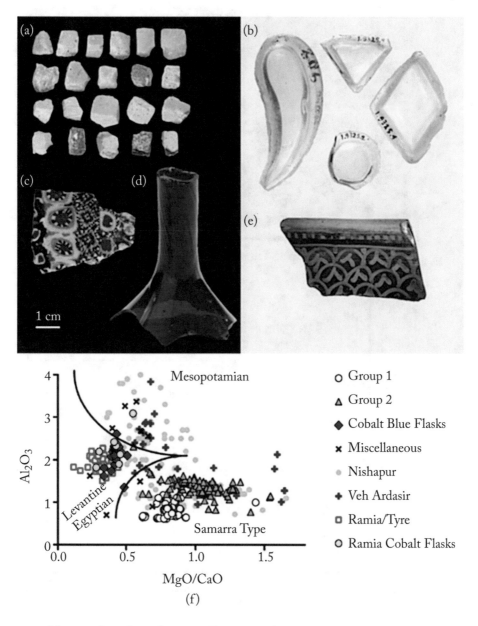

Figure 2.29: Glass artifacts from Samarra. Regularly shaped mosaic tesserae glass (a), glass inlays of plant ash group 1 (b), fragment of millefiori glass tile of plant ash group 2 (c), cobalt blue flask neck (d), rim fragment of painted glass bowl belonging to the miscellaneous samples (e), and Al_2O_3 vs. MgO to CaO ratios determined by laser ablation ICP-MS (f). Reprinted from [29].

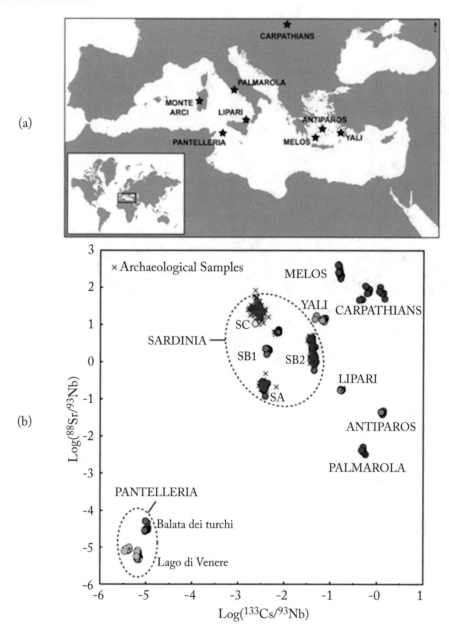

Figure 2.30: Map of the main obsidian sources in the Mediterranean area (a) and comparison of the log (^{88}Sr/^{93}Nb) ratios obtained by ICP-MS on known geological samples from different Mediterranean regions and Neolithic archaeological samples from the Tyrrhenian area (b). Reprinted from [30].

2.11   REFERENCES

[1] M. Mantler and M. Schreiner, X-ray fluorescence spectrometry in art and archaeology, *X-Ray Spectrom.*, 29:3–17, 2000. DOI: 10.1002/(sici)1097-4539(200001/02)29:1%3C3::aid-xrs398%3E3.0.co;2-o. 94

[2] A. Macková, D. Macgregor, F. Azaiez, J. Nyberg, and E. Piasetzky, Nuclear physics for cultural heritage a topical review by the Nuclear Physics Division of the European Physical Society, Edited by Edp Sciences, 2016. 95

[3] L. Burgio, R. J. H. Clark, and R. R. Hark, Raman microscopy and x-ray fluorescence analysis of pigments on medieval and renaissance Italian manuscript cuttings, *Proc. Natl. Acad. Sci.*, 107:5726–5731, 2010. DOI: 10.1073/pnas.0914797107. 96

[4] K. V. Manukyan, B. J. Guerin, E. J. Steck, A. Aprahamian, M. Wiescher, D. T. Gura, and Z. D. Schultz, Multiscale X-ray fluorescence mapping complemented by Raman spectroscopy for pigment analysis of a 15th century Breton manuscript, *Anal. Meth.*, 8:7696–7701, 2016. DOI: 10.1039/c6ay02301k. 97, 98, 119

[5] D. L. Howard, M. D. De Jonge, D. Lau, D. Hay, M. Varcoe-Cocks, C. G. Ryan, R. Kirkham, G. Moorhead, D. Paterson, and D. Thurrowgood, High-definition X-ray fluorescence elemental mapping of paintings, *Anal. Chem.*, 84:3278–3286, 2012. DOI: 10.1021/ac203462h. 98

[6] J. Dik, K. Janssens, G. Van Der Snickt, L. Van Der Loeff, K. Rickers, and M. Cotte, Visualization of a lost painting by Vincent van Gogh using synchrotron radiation based X-ray fluorescence elemental mapping, *Anal. Chem.*, 80:6436–6442, 2008. DOI: 10.1021/ac800965g. 98, 99

[7] E. Brun, M. Cotte, J. Wright, M. Ruat, P. Tack, L. Vincze, C. Ferrero, D. Delattre, and V. Mocella, Revealing metallic ink in Herculaneum papyri, *Proc. Natl. Acad. Sci.*, 113:3751–3754, 2016. DOI: 10.1073/pnas.1519958113. 98, 100

[8] M. A. Rizzutto, M. V. Moro, T. F. Silva, G. F. Trindade, N. Added, M. H. Tabacniks, E. M. Kajiya, P. H. V. Campos, A. G. Magalhães, and M. Barbosa, External-PIXE analysis for the study of pigments from a painting from the Museum of Contemporary Art, *Nucl. Instrum. Meth. Phys. Res. Sect. B*, 332:411–414, 2014. DOI: 10.1016/j.nimb.2014.02.108. 102, 103

[9] N. Grassi, P. Bonanni, C. Mazzotta, A. Migliori, and P. A. Mand, PIXE analysis of a painting by Giorgio Vasari, *X-Ray Spectrom.*, 38:301–307, 2009. DOI: 10.1002/xrs.1181. 102, 104

[10] T. Calligaro, V. Gonzalez, and L. Pichon, PIXE analysis of historical paintings: Is the gain worth the risk? *Nucl. Instrum. Meth. Phys. Res. Sect. B*, 363:135–143, 2015. DOI: 10.1016/j.nimb.2015.08.072. 104

[11] K. Manukyan, C. Fasano, A. Majumdar, G. F. Peaslee, M. Raddell, E. Stech, and M. Wiescher, Surface manipulation techniques of Roman Denarii, *Appl. Surf. Sci.*, 493:818–828, 2019. DOI: 10.1016/j.apsusc.2019.06.296. 104, 106

[12] F. J. Ager, A. I. Moreno-Suárez, S. Scrivano, I. Ortega-Feliu, B. Gómez-Tubío, and M. A. Respaldiza, Silver surface enrichment in ancient coins studied by micro-PIXE, *Nucl. Instrum. Meth Phys. Res. Sect. B*, 306:241–244, 2013. DOI: 10.1016/j.nimb.2012.12.037. 107, 108

[13] X. Li, X. Wu, Y. Zhao, Q. Wen, Z. Xie, Y. Yuan, T. Tong, X. Shen, and H. Tong, Composition/structure and lacquering craft analysis of Wenzhou Song dynasty lacquerware, *Anal. Meth.*, 8:6529–6536, 2016. DOI: 10.1039/c6ay01694d. 109

[14] C. M. A. McQueen, D. Tamburini, and S. Braovac, Identification of inorganic compounds in composite alum-treated wooden artefacts from the Oseberg collection, *Sci. Rep.*, 8:1–8, 2018. DOI: 10.1038/s41598-018-21314-z. 109, 110

[15] L. Beck, H. Rouselière, J. Castaing, A. Duran, M. Lebon, B. Moignard, and F. Plassard, First use of portable system coupling X-ray diffraction and X-ray fluorescence for in-situ analysis of prehistoric rock art, *Talanta*, 129:459–464, 2014. DOI: 10.1016/j.talanta.2014.04.043. 112, 113

[16] E. Possenti, C. Colombo, C. Conti, L. Gigli, M. Merlini, J. R. Plaisier, M. Realini, and G. D. Gatta, What's underneath? A non-destructive depth profile of painted stratigraphies by synchrotron grazing incidence X-ray diffraction, *Analyst*, 143:4290–4297, 2018. DOI: 10.1039/c8an00901e. 112

[17] A. Duran, A. López-Montes, J. Castaing, and T. Espejo, Analysis of a royal 15th century illuminated parchment using a portable XRF-XRD system and micro-invasive techniques, *J. Archaeol. Sci.*, 45:52–58, 2014. DOI: 10.1016/j.jas.2014.02.011. 112, 114

[18] E. Welcomme, P. Walter, P. Bleuet, J. L. Hodeau, E. Dooryhee, P. Martinetto, and M. Menu, Classification of lead white pigments using synchrotron radiation micro X-ray diffraction, *Appl. Phys. A Mater. Sci. Process*, 89:825–832, 2007. DOI: 10.1007/s00339-007-4217-0. 113, 115

[19] L. Burgio and R. J. H. Clark, Library of FT-Raman spectra of pigments, minerals, pigment media and varnishes, and supplement to existing library of Raman spectra of pigments with visible excitation, *Spectrochim. Acta A*, 57:1491–1521, 2001. DOI: 10.1016/s1386-1425(00)00495-9. 117

[20] N. C. Scherrer, Z. Stefan, D. Francoise, F. Annette, and K. Renate, Synthetic organic pigments of the 20th and 21st century relevant to artist's paints: Raman spectra reference collection, *Spectrochim. Acta A*, 73:505–524, 2009. DOI: 10.1016/j.saa.2008.11.029.

[21] G. Marucci, A. Beeby, A. W. Parker, and C. E. Nicholson, Raman spectroscopic library of medieval pigments collected with five different wavelengths for investigation of illuminated manuscripts, *Anal. Meth.*, 10:1219–1236, 2018. DOI: 10.1039/c8ay00016f. 117

[22] M. Sendova, V. Zhelyaskov, M. Scalera, and M. Ramsey, Micro-Raman spectroscopic study of pottery fragments from the Lapatsa Tomb, Cyprus, CA 2500 BC, *J. Raman Spectrosc.*, 36:829–833, 2005. DOI: 10.1002/jrs.1371. 117, 118

[23] K. Manukyan, M. Raddell, E. Sestak, D. Gura, Z. Schultz, and M. Wiescher, Pigment and ink analysis of medieval books through complementary spectroscopy techniques, *Glob. J. Archaeol. Anthropol.*, 3(555619):1–4, 2018. DOI: 10.19080/GJAA.2018.03.555619. 119

[24] L. Monico, G. Van Der Snickt, K. Janssens, W. De Nolf, C. Miliani, J. Dik, M. Radepont, E. Hendriks, M. Geldof, and M. Cotte, Degradation process of lead chromate in paintings by Vincent van Gogh studied by means of synchrotron X-ray spectromicroscopy and related methods, 2. Original paint layer samples, *Anal. Chem.*, 83:1224–1231, 2011. DOI: 10.1021/ac1025122. 120, 121

[25] M. Leona, Microanalysis of organic pigments and glazes in polychrome works of art by surface-enhanced resonance Raman scattering, *Proc. Natl. Acad. Sci.*, 106:14757–14762, 2009. DOI: 10.1073/pnas.0906995106. 120, 122

[26] A. Nevin, J. L. Melia, I. Osticioli, G. Gautier, and M. P. Colombini, The identification of copper oxalates in a 16th century Cypriot exterior wall painting using micro FTIR, micro Raman spectroscopy and gas chromatography-mass spectrometry, *J. Cult. Herit.*, 9:154–161, 2008. DOI: 10.1016/j.culher.2007.10.002. 125, 126

[27] S. Švarcová, E. Kočí, J. Plocek, A. Zhankina, J. Hradilová, and P. Bezdička, Saponification in egg yolk-based tempera paintings with lead-tin yellow type I, *J. Cult. Herit.*, 38:8–19, 2019. DOI: 10.1016/j.culher.2018.12.004. 126, 128

[28] B. Giussani, D. Monticelli, and L. Rampazzi, Role of laser ablation-inductively coupled plasma-mass spectrometry in cultural heritage research: A review, *Anal. Chim. Acta.*, 635:6–21, 2009. DOI: 10.1016/j.aca.2008.12.040. 130

[29] N. Schibille, A. Meek, M. T. Wypyski, J. Kröger, M. Rosser-Owen, and R. W. Haddon, The glass walls of Samarra (Iraq): 19th-century Abbasid glass production and imports, *PLoS One*, 13:1–125, 2018. DOI: 10.1371/journal.pone.0201749. 130, 131

[30] M. Orange, F. X. Le Bourdonnec, A. Scheffers, and R. Joannes-Boyau, Sourcing obsidian: A new optimized LA-ICP-MS protocol, *Sci. Technol. Archaeol. Res.*, 2:192–202, 2016. DOI: 10.1080/20548923.2016.1236516. 130, 132

CHAPTER 3

Isotope Analysis Techniques

3.1 INTRODUCTION

It will doubtless interest readers of Nature to know that other elements besides neon have now been analyzed in the positive-ray spectrograph with remarkable results.

(Francis W. Aston, The Constitution of the Elements, 1919)

This chapter is concerned with the analysis of neutron induced radioactive isotopes as well as the analysis of stable isotope distribution in cultural heritage materials. Neutron Activation Analysis (NAA) is a well-established and widely accepted technique with a broad range of applications in art forensics, archaeology, and anthropology. In Section 1.6, we discussed the underlying physics of activating materials through charged particle or neutron bombardment, generating characteristic decay radiation that can be used for material analysis purposes. While charged-particle-based activation is primarily used for producing long-lived radioactive materials for medical or industrial applications, the following Section 3.2 concentrates on neutron activation analysis as one of the main nuclear physics tools in the analysis of historical artifacts.

Isotope distribution in cultural heritage materials often depends on the geological and chemical history of the sample and is affected by chemical fractionation processes that are characteristic for the specific location. In other cases, the decay of long-lived radioactive isotopes also influences the associated abundance distribution through the build-up of daughter nuclei over long-term geological periods, providing information about the provenance and origin of archaeological and anthropological samples. A number of examples for applications of these methods will be presented in Section 3.3.

3.2 NEUTRON ACTIVATION TECHNIQUE

Activation analysis is based on the synthetic production of radioactivity in the sample, in order to use decay characteristics—ranging from characteristic decay radiation to characteristic time decay—as a unique analytical signature for identifying elemental or isotopic components in the sample. The neutron activation technique is an alternative or complementary technique to X-ray based methods, which rely on the analysis of transitions in the atomic nucleus. Neutron activation provides not only information about elemental abundances but also provides useful isotope analysis. In addition, neutrons penetrate much deeper into the sample than charged particles or X-rays and are therefore extremely useful for probing the inner layers of materials.

3.2.1   NEUTRON SOURCES

Neutrons have a short lifetime of just under 15 minutes. This necessitates the implementation of neutron sources for the efficient activation of materials. The simplest cases of neutron sources are a mix of heavy α emitting actinides and the beryllium isotope ^{9}Be. The most common α emitters used in neutron sources include ^{241}Am, ^{238}Pu, ^{239}Pu, ^{210}Po, and ^{226}Ra, with ^{241}Am being the most common. The α particles are produced by the natural decay of these actinides at fairly high energies and the particles are rapidly captured by beryllium, in the ^{9}Be$(\alpha, n)^{12}$C reaction. Beryllium is the most common low Z material used in alpha-neutron sources because of its relatively high neutron yield, but other materials such as fluorine, lithium, and boron can also be used. The beryllium reaction has a very high cross section, converting most of the emitted α particles into neutrons.

The strength of such a source is determined by the activity of the alpha emitter. Activities of 0.5–40 Ci (18.5 GBq–1.48 TBq) are common, although portable density gauges might employ 10–50 mCi (0.37–1.85 GBq) sources. An Am-Be source typically generates a neutron yield of ca. 2.0–$2.4 \cdot 10^6$ neutrons/sec or 5.4–$6.5 \cdot 10^4$ neutrons/sec per GBq. The Am-Be neutron source is typically embedded in a steel container with internal cadmium or boron absorber material, with space for the released helium that is not captured by beryllium.

Alternative neutron sources are the fission sources, which are based on the release of free neutrons in the natural fission process of actinide isotopes, such as ^{238}Pu, ^{242,244}Cm, and ^{252}Cf. These actinide isotopes fission naturally, releasing 3–4 neutrons per fission event. The over-all activity of the fission source characterizes the neutron flux. The high neutron yield coming from ^{252}Cf, with a half-life of 2.645 years, permits the construction of physically small neutron sources. The released neutron yield in units of weight or source activity is 2.3–$2.4 \cdot 10^{12}$ n, s^{-1}, g^{-1} or $4.4 \cdot 10^9$ n, s^{-1}, Ci^{-1} or $1.2 \cdot 10^8$ n, s^{-1}, GBq^{-1}. These types of sources are to a certain extent portable and therefore very convenient for quick applications of the technique. However, they require the production of the radioactive actinide species within a nuclear reactor, and this activity is often limited, particularly when used in non-governmental regulated settings. There-fore, the neutron flux remains limited.

To achieve higher neutron flux conditions, either nuclear reactors themselves must serve as the neutron source or nuclear accelerators can be used to produce the nuclear reactions that provide a large enough neutron yield. Both of these methods have been described in Chapter 1. Nuclear reactors produce neutrons by fission of the reactor fuel, typically either ^{235}U or ^{239}Pu. High-flux reactor sources provide a neutron flux of up to 10^{12} to 10^{15} n cm^{-2} s^{-1}. These neutrons are initially emitted at high energies during the fission process but scatter on the surrounding atoms, distributing and loosing energy until they cool down to energies corresponding to the temperature of the reactor core. This process is called thermalizing. The neutron flux can be modified by several orders of magnitude, depending on the reactor operation and the location of the activation station. Research reactors, as opposed to power reactors, provide the possibility of positioning samples close to the reactor core and can maximize the neutron flux by incorporating beam ports that allow neutrons to exit from the reactor core and be transported through neutron guidelines to the experimental station. This reduces the actual neutron flux by about two orders of magnitude compared to the flux in the core. In any case, the neutron flux is substantially higher than the flux provided by a portable neutron source as described above.

Accelerators offer a broad range of possibilities for producing a substantial neutron flux. Electron accelerators produce an intense bremsstrahlung spectrum by bombarding a target with high Z materials such as tungsten. The bremsstrahlung is then used to produce neutrons via (γ,n) photon production processes on light target material, such as deuterium or beryllium, with low neutron binding energy. With the increasing energy of the electron beam, the energy of the bremsstrahlung spectrum increases, producing higher amounts of the energy neutrons. In some cases, neutron energies are obtained that are comparable to the energy of the bombarding electrons. An alternative is the use of positive ion accelerators or neutron generators that accelerate protons, deuterons, or α particles into a low Z target, triggering nuclear reactions such as $^3\mathrm{H}(^2\mathrm{H}, n)^4\mathrm{He}$—the so-called $d - t$ reaction that fusions a deuterium and a tritium nucleus to helium, releasing neutrons at an energy of 14 MeV. Other possible reactions are the $d - d$ reaction where two deuterons fusion via $^2\mathrm{H}(^2\mathrm{H}, n)^3\mathrm{He}$ to helium-3, releasing neutrons at an energy of 2.5 MeV. Frequently used is the proton induced reaction on lithium, $^7\mathrm{Li}(p, n)^7\mathrm{Be}$, which produces neutrons at an energy above 2 MeV proton energy. Neutron energy increases with proton energy, making this a very versatile source for applications. Finally, there are α induced reactions on light target material, as with the aforementioned beryllium, $^9\mathrm{Be}(\alpha, n)^{12}\mathrm{C}$, the difference being that the energetic α particles are produced by an accelerator system rather than by a nuclear decay process. Accelerator-based neutron sources can be turned off. This is a major advantage over decay, fission, or reactor sources since it reduces safety risks and minimizes unwanted activation processes in the source environment.

3.2.2   SMALL SAMPLES ACTIVATION

Many applications of neutron activation are targeted for the analysis of small samples, ranging from ancient coin material to pottery shards and human remnants. These samples are small and can be easily positioned in a neutron environment for activation. The actual activation time, during which the sample is exposed to the neutrons, depends on the half-life of the activated isotopes; to reach maximum activity in the sample, the required activation time is 3–5 times the half-life of the activated isotope. However, in many cases, the half-life is too long and the exposure periods will be considerably shorter than recommended.

A perfect application of the neutron activation technique is the analysis of historic silver or gold coins in order to determine the actual content of noble metals in the coin, for exploring coin purity and manufacturing techniques. Figure 3.1 shows a medieval German coin, the Bonner Großpfennig (large penny), minted in the archdiocese of Cologne in the Late Middle Ages. The Großpfennig represented the transition from the penny currency of the Early Middle Ages to the Groschen coinage that began in Germany in the 14th century and reflected to a certain

Figure 3.1: Bonner Großpfennig minted by the archbishop of Cologne, Heinrich von Virneburg (1275–1297), in a weight that corresponded to the weight of a Roman Denarius.

extent the inflation in that period of time, which required the introduction of a new coinage system.

Neutron capture would occur on the silver content via the $^{109}Ag(n, \gamma)^{110}Ag$ reaction [1]. ^{110}Ag has a half-life of 250 days and emits a characteristic γ radiation of 658 keV; gold would be activated via the $^{197}Au(n, \gamma)^{198}Au$ reaction that decays with a half-life of only 2.5 days, emitting γ radiation of 412 keV. Possible copper content in the coin, either from contamination in the initial ore melting process or as targeted dilution of the silver content, can be measured by activation through the $^{63}Cu(n, \gamma)^{64}Cu$ reaction. ^{64}Cu has a half-life of 12,700 hours and emits a characteristic γ decay radiation of 1346 keV. Figure 3.2a shows a typical γ-spectrum obtained by neutron capture on the Großpfennig.

In Figure 3.2b, analysis of 30 Großpfennig samples shows that the silver content ranges between 97 and 98%, which exceeds the quality of Sterling silver (92.5%). Spurious amounts of gold appear at a level of 0.1–0.2%. Gold and silver are chemically similar and therefore hard to separate in analysis. Silver coins were typically alloyed with copper. The copper content of these coins ranges between 2 and 3%. The only exception is coin #12 with a copper content of 20% and a reduced silver content of 80% accordingly.

Besides coins, other materials such as shards from pottery can be tested with neutron activation techniques. Since the elemental abundance distribution in the original clay material depends on the original location, neutron activation analysis is helpful for determining the provenance of the pottery material. Neutron activation is used as a standard tool in the analysis of ancient pottery worldwide, ranging from Mesoamerican to Mesopotamian pottery, to identify the origin of the material. Since pottery in the ancient world was frequently used as container

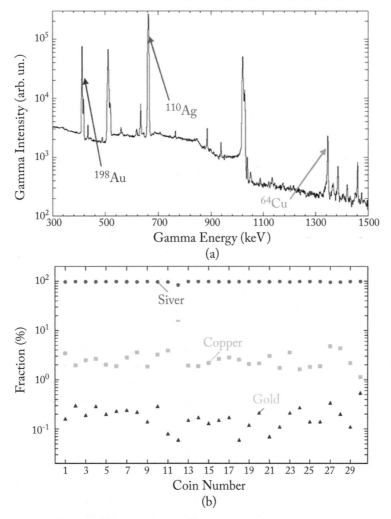

Figure 3.2: Typical γ spectrum of a coin after neutron activation. Highlighted are the character-istic decay lines for gold, silver, and copper (a). The fraction of silver, copper, and gold content in a number of Bonner Großpfennig coins. One can clearly see that the silver content of the coins is constant except for coin #12, which shows with a 20% higher enrichment in copper (b). (The fractions are shown on a logarithmic scale.)

material, these techniques help to generate information about trade and trading routes. One of the most important examples was the neutron activation analysis of the pottery vessels contain-ing the ancient Qumran scrolls, which were found in the late 1940s in the caves located near the Dead Sea in Palestine [2]. The elemental composition of the heavy elements Ce, Co, Cs,

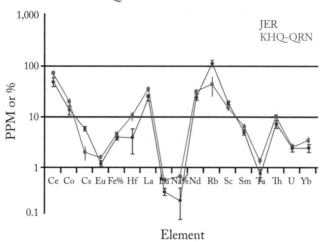

Figure 3.3: Comparison of the elemental composition of pottery from Qumran (KHQQRN) and Jerusalem (JER) shows a clear correlation in the characteristic abundance distribution of the heavy elements. Figure plotted from the data presented in [2].

Eu, Fe, Hf, La, Lu, Na, as well as Nd, Rb, Sc, Sm, Ta, Th, U, and Yb (in alphabetical order) in the Qumran samples, as well as samples from pottery produced in Jerusalem at the time, are shown in Figure 3.3. The close correlation showed that at least this sample of Qumran pottery originated in Jerusalem, indicating contact between religious communities at Qumran and Jerusalem.

3.2.3   NEUTRON-AUTORADIOGRAPHY

Neutron-autoradiography has emerged as an important tool in the analysis of paintings. While neutron activation primarily serves for identifying the chemical composition of pigments by their characteristic γ-emission, the autoradiograph helps in the analysis of painting techniques, such as brush strokes and underpaintings that are often well hidden in the artwork [3]. This analysis requires neutron irradiation of the entire painting using a homogeneous neutron flux, which can be complemented by point-by-point raster activation measurements. Such methods have been developed at the Hahn–Meitner Institute in Berlin using a small reactor with an external neutron guide to achieve a neutron flux of about 10^{12}–10^{14} neutrons/cm^2/s at the activation station (Figure 1.52).

The reactor neutrons typically have thermal energies, which, according to the $1/v$ law, increase the probability of, or cross section to undergo, neutron capture reactions with the nuclei of the pigments that the painters have used. Neutron capture on neutron rich isotopes such as ^{31}P$(n, \gamma)^{32}$P or ^{53}Mn$(n, \gamma)^{54}$Mn produces long-lived radioactive isotopes, which decay with a characteristic half-life of $T_{1/2} = 14$ days and $T_{1/2} = 2.6$ hours, respectively. The transformation

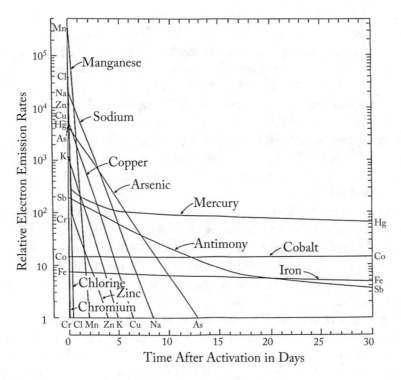

Figure 3.4: Relative rates of β emission during the radioactive decay of neutron-activated pigments with the painting *Saint Rosalie Interceding for the Plague-stricken of Palermo* https://www.metmuseum.org/art/metpublications/Art_and_Autoradiography_Insights_into_the_Genesis_of_Paintings_by_Rembrandt_Van_Dyck_and_Vermeer.

of specific isotopes in the pigments into radioactive species causes a general activation of the painting. The subsequent radioactive decay of each of the activated isotopes follows the decay law and, depending on the initial neutron exposure, the emitted decay radiation is sufficient to expose photographic film placed in direct contact with the paintings. At different times, different decay patterns will emerge in the radiograph, depending on which radioactive decay process dominates. In the here discussed specific case, the radiation from ^{54}Mn in umbra or dark ochre pigments will expose the photographic film during the first 5–7 hours after neutron exposure, and decay radiation from ^{32}P in bone-black will dominate the exposure of the film many days after the ^{56}Mn has decayed away. Figure 3.4 shows the decay curves of several radioactive isotopes of elements that can typically be found in pigments of ancient oil paintings.

The resulting autoradiographs reveal structural details in depth, of both the painting and its support. Analysis of such a series of autoradiographs permits the identification of the pigments used in a painting. Analysis also provides information about the manner in which the pigments were originally laid down by the artist, reflecting characteristic layer techniques to generate depth. Pigment distribution throughout the body of the painting is also revealed, indicating the way the artist generated shades and tones by admixtures. The stability of oil paintings is unaffected by the activation procedure. The neutron radiation dose absorbed by a painting is of the order of 50 rads (0.5 Gy). A substantially lower flux would reduce the intensity of the emitted decay radiation, leading to a reduction in the exposure of the radiographs. Paintings subjected to absorbed doses 100 times as great as the dose necessary for autoradiography exhibit no changes in color, hardness, flexibility and solubility three years after their original activation. Nevertheless, as indicated by Figure 3.4, some of the radioactive elements produced by neutron activation have long half-lives lasting for longer periods of time. Yet the associated activity is well within permissible levels.

A number of interesting cases can be found in the literature, [4] but we will present only one example, a painting of Saint Rosalie praying for the plague-stricken city of Palermo in 1624 by Anthony van Dyke, to illustrate the method in more detail (Figure 3.5). While Figure 3.5a shows a direct optical image of the painting, Figure 3.5b shows the difference of an X-ray radiography image of the same painting, with neutron autoradiography images taken at different times in Figures 3.5c–d.

The X-ray radiograph reflects the absorption in the heavy wood frame of the canvas as well as in the lead-containing pigments, such as lead white which is a complex salt, $2 PbCO_3 \cdot Pb(OH)_2$, containing both lead carbonate and hydroxide molecules (Figure 3.5b). The autoradiograph, taken 4–5 hours after neutron irradiation (Figure 3.5c), shows primarily the decay radiation of ^{54}Mn, which is in the umbra pigment $Fe_2O_3 + MnO$. The white section in the center of the painting indicates a repair with modern organic pigment materials, which are not being activated. An autoradiograph, taken 2–4 days after neutron irradiation (Figure 3.5d), shows primarily the decay radiation of ^{64}Cu with a half-life of $T_{1/2} = 12.8$ h; copper is a component of the blue pigment azurite $CuCO_3 Cu(OH)_2$, which was used for the bluish gray painting of sky and clouds. In addition to azurite, Van Dyke used a bit of ultramarine, the finest and most expensive blue used by painters. Ultramarine is a sodium aluminosilicate ($Na_{8-10}Al_6Si_6O_{24}S_{2-4}$), which is reflected in the decay radiation of ^{24}Na that decays with a half-life of $T_{1/2} = 15$ h. Other radioactive isotopes produced by neutron activation are just too short-lived to be detected in the autoradiography technique. The eighth autoradiograph, 8–20 days after irradiation (Figure 3.5e), reveals an underpainting (here rotated by 180°), executed in bone black, a pigment, $C+Ca_3(PO_4)_2$ that is typically produced from bone ash and was used for sketches. The exposure comes from the decay radiation of ^{32}P with a half-life of $T_{1/2} = 14.3$ days. The image clearly resembles a self-portrait by Anthony van Dyke of 1622 (Figure 3.5f). This indicates that Van Dyke, after receiving the commission for painting Saint Rosalie, simply used the canvas with his not fully executed self-portrait.

A further example is the portrait of the *Man with the Golden Helmet*, which supposedly was an oil painting by the Dutch painter Rembrandt van Rijn (Figure 3.6a). There had been

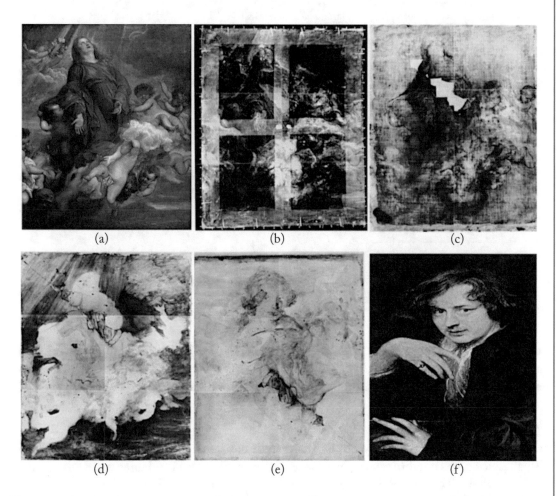

Figure 3.5: *Saint Rosalie Interceding for the Plague-stricken of Palermo* (1624) by Anthony van Dyke (a). Shown are an X-ray radiograph (b), an autoradiograph, 4–5 hours after neutron irradiation (c), an autoradiograph, 2–4 days after neutron irradiation (d), and the eighth autoradiograph, 8–20 days after irradiation (e), revealing a sketch of a young man, presumably an uncompleted self-portrait (here rotated by 180°). For comparison a known self-portrait of Anthony van Dyke painted in 1622 (f) https://www.metmuseum.org/art/metpublications/Art_and_Autoradiography_Insights_into_the_Genesis_of_Paintings_by_Rembrandt_Van_Dyck_and_Vermeer.

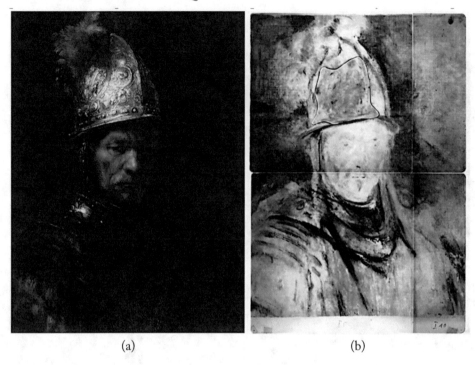

(a)   (b)

Figure 3.6: Oil painting of the *Man with the Gold Helmet* (a). Autoradiograph reflecting the activity of ^{32}P with a half-life of $T_{1/2} = 14.3$ days (b). The early sketch of the figure clearly reveals a different paint stroke technique than used by Rembrandt in his other paintings.

suspicions that this was actually not by Rembrandt but rather, a work by one of his students. This suspicion was re-affirmed by a neutron activation analysis, which demonstrated that, unlike Rembrandt in all of his paintings, the painter did not use cinnabar or vermillion, a pronounced red pigment HgS containing mercury. The use of vermillion would have been reflected after approximately 30 days in the decay radiation of ^{203}Hg with a half-life of $T_{1/2} = 48$ days. There was no indication of late-term exposure of the photographic film at such a late time after neutron irradiation [5]. Another indication was the brushstroke indicated by the ^{32}P analysis $T_{1/2} = 14.3$ days in an early autoradiography (Figure 3.6b). It reflected a different brush technique than observed in well-known and documented Rembrandt paintings. These results indicated that this painting was not painted by Rembrandt but at best can be assigned to one of his students.

With respect to the actual painter of this famous portrait, the similarity in the design of the actual gold helmet with helmets manufactured at that time in the Fugger armories of the city of Augsburg, as well as the specific painting style, points to Johann Ulrich Mayr (1629–1704), who was born and lived in Augsburg. He was a gifted portrait painter, who had studied at the

Rembrandt school in 1648/49. His paintings frequently reflect an attachment to weapons and armory.

3.3   ISOTOPE ANALYSIS TECHNIQUES

This section is concerned with the analysis of stable isotope abundance distribution in objects that have emerged through chemical and physics fractionation processes. Looking at a typical nuclide chart, as discussed in Chapter 1, one finds the abundances of the stable isotopes listed as fixed values. For example, in carbon, the two stable isotopes ^{12}C and ^{13}C have a fraction of 98.9% and 1.1%, respectively. There is a third isotope ^{14}C, which is radioactive with a half-life of 5,730 years and is continuously produced by cosmic ray interaction in the atmosphere via the $^{14}N(n,p)^{14}C$ reaction, as outlined in Section 1.8.1. ^{14}C, therefore, occurs at a very small equilibrium level of about 10^{-10}% depending on many geophysical aspects, ranging from the solar activity level to the strength of the earth's magnetic field. Carbon-14 is used in dating historical artifacts and as a tracer of biochemical processes, as discussed in Chapter 5. Another important element is oxygen, with three stable isotopes ^{16}O, ^{17}O, and ^{18}O that carry solar abundances of 99.762%, 0.0373%, 0.2%, respectively. Each element has both stable and radioactive isotopes, which appear in certain fractions up to lead (Pb) with four stable isotopes ^{204}Pb, ^{206}Pb, ^{207}Pb, and ^{208}Pb that occur in fractions of 1.4%, 24.1%, 22.1%, and 52.4%, respectively. The only elements without any stable isotope components are technetium and the noble gas radon, but radioactive components for both of these are observed on earth as a result of natural radioactive decay patterns. These isotope ratios are typically called solar isotope ratios since they are averaged over all observations in the solar system [6]. They have been caused by nuclear reaction processes in star generations prior to the origin of the solar system and are imbedded in the original solar material [7].

Isotope ratios for specific elements can change locally due to geological, climatological, biological, chemical, and physical fractionation processes during the history of the earth. For the ^{13}C to ^{12}C ratio, the fractioning mechanisms are primarily based on the photosynthesis of plants, converting CO_2 into O_2 with the carbon converted to sugar as plant nutrient. This process naturally enhances the lighter carbon isotopes, so that the carbon in plants is depleted in heavier carbon isotopes such as ^{13}C and ^{14}C. This fractionation occurs in enzymatic conversion processes, the so-called carboxylation reaction, that cause up to 28% depletion in 1 [8]. After photosynthesis, the isotope ^{13}C is depleted on average by 1.8% in comparison to its initial ratio in the atmosphere.

Another important example for anthropological considerations is the ratio of ^{18}O to ^{16}O isotopes, as reflected in human teeth, hair, and bones. In rainy coastal regions, the ^{18}O content is enhanced, since water molecules H_2O that include ^{18}O are slightly heavier than the ones containing ^{16}O and therefore condensate and fall more easily down with rain. The cloud becomes gradually depleted in ^{18}O, while drifting inland and so, rainwater far off the coast contains a lower ^{18}O content, as can be easily measured with high precision mass analyzers [9]. The oxygen isotope ratio in human body parts can be measured in great detail, tracing the location of the individuals from birth to death. Tooth enamel, for example, is formed in early childhood, and the isotope ratio would point to the origin of the individual. In dentin and bone, the ratio changes with time, which allows for tracing the person's location during his lifetime and death.

A third example of isotope change is the isotope distribution in lead material, which was used, in particular, in ancient Greek and Roman times for water pipes, aqueducts, tank linings, plates, and cooking pots but also for early cosmetics, paints, and pigments in lead-rich glazes. Lead was an important trading object, found in shipwrecks all over the Mediterranean. Lead was also an important product of the ancient mining industry, but the fraction of specific lead isotopes primarily depends on the geological composition of the mining ore, in particular their content of long-lived radioactive isotopes such as ^{232}Th, ^{235}U, and ^{238}U, which are the mother elements of the natural radioactive decay chains. The natural decay chains are a series of radioactive decay processes that transform the so-called mother isotopes, ^{232}Th, ^{235}U, and ^{238}U to Pb, causing a selective feeding of the lead isotopes ^{208}Pb, ^{207}Pb, and ^{206}Pb, respectively, through their specific sequential series of transformations. A fourth decay chain of ^{237}Np, passing through radioactive ^{209}Pb, ends with the production of the extremely long-lived ^{209}Bi ($T_{1/2} = 1.9 \cdot 1019$ years), which eventually decays by α emission to ^{205}Tl. Depending on the initial thorium, uranium, and neptunium composition and the geological age of the original mineral material, different isotopic abundances characterize the isotope distribution in lead that is being measured in archaeological samples today.

There are many more of these chemical and physical fractionation processes, for heavier elements ranging from rubidium (Rb, $A = 37$) to silver (Ag, $A = 47$), and mercury (Hg, $A = 80$), which are primarily used for tracing the origin of materials or for identifying ancient paths of trading and commerce, leading to a deeper understanding of the ancient world. A small selection of examples will be discussed in the following section.

3.3.1   MASS ANALYSIS AND MASS ANALYZERS

Deviations from the solar abundances due to chemical and physical fractionation processes are in many cases miniscule. To identify the differences in the abundances of isotopes with slightly different nuclear masses, high-precision mass analysis is required for reliable measurement. In the following section, we therefore want to briefly introduce the standard types of mass analyzers and the importance of standard samples, which are necessary for normalizing the experimental results when documenting isotope ratios.

The typical approach in mass analysis is based on the fact that accelerated charged particles in a magnetic field are deflected by the Lorentz force. Their charge over mass ratio q/m depends on the radius R of their deflection, the strength of the magnetic field B perpendicular to their trajectory, and their velocity v when entering the magnetic field:

$$\frac{q}{m} = \frac{B \cdot R}{v}. \tag{3.1}$$

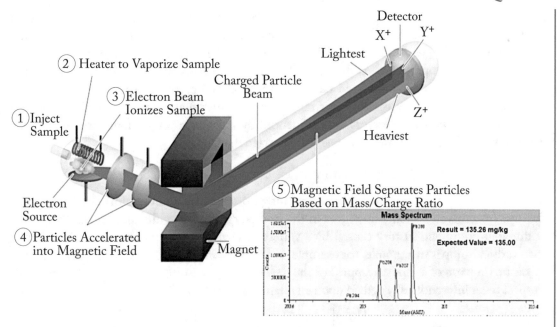

Figure 3.7: Modern mass analyzer, with lead ions generated by vaporization and ionization in a plasma source before being accelerated toward the magnetic field and deflected depending on their specific masses. The insert shows the spectrum for different lead isotopes as a function of their mass. The peak height or the integrated peak area corresponds to the specific lead isotope abundance and the ratio of peak areas yield the abundance ratios between different lead isotopes (http://www.chem.ucalgary.ca/courses/350/Carey5th/Ch13/ch13-ms.html).

Highest resolution for mass separation can be obtained by a well-defined radius in the magnetic field. Figure 3.7 shows a sketch of a modern mass analyzing system. Atoms are ejected from a small sample, through vaporization or sputtering by cesium ions or high intensity lasers, and ionized. The details of this process depend on the kind of ion source being used in the instrument. Positively charged ions with charge q are accelerated in an electric potential V to a kinetic energy of: $E = q \cdot V = \frac{1}{2} m \cdot v^2$, which corresponds to a velocity of: $v = \sqrt{\frac{2 \cdot q \cdot V}{m}}$ for the ions entering the perpendicular magnetic field B. The radii R_1 and R_2 of the orbital trajectories of two charged particles in a magnetic field scale with the particle masses m_1 and m_2, respectively:

$$\frac{m_1}{m_2} = \frac{\left(\frac{q \cdot B}{v}\right)}{\left(\frac{q \cdot B}{v}\right)} \frac{R_1}{R_2} = \frac{R_1}{R_2}. \tag{3.2}$$

According to this equation, two lead ions with mass $m_1 = 204$ amu and $m_2 = 208$ amu (amu is the atomic mass unit as defined in Chapter 1) would scale to two different orbits of $R_2 = 1$ m and $R_1 = 0.98$ m, respectively. A similar exercise can be done for oxygen ions of $m_1 = 16$ amu and $m_2 = 17$ amu, with R_2 being 1 m again, while R_1 would be 0.94 m. Each trajectory would correspond to a different exit point of the particle from the magnetic field range, where it can be detected using appropriate ion detectors. Typically, a small aperture is positioned behind the magnetic field area, defining a fixed radius for the analyzed particles. Ramping the magnetic field allows for covering a certain mass range. The size of the aperture defines the resolution of the spectrum $m/\Delta m$, which is also influenced by instrument dependent variations in the accelerating voltage V and the hysteresis of the magnetic field.

3.3.2   ISOTOPE FRACTIONATION

Mass analyzers are a powerful tool in determining isotope ratios and therefore also in mapping isotope fractionation, small differences in the isotope ratios of specific samples which indicate a different origin and history caused by a variety of environmental impacts. Fractionation of two carbon isotopes in a sample, for example, is expressed in terms of $\delta^{13}C$. This parameter is a measure (in parts of a thousand ppm) of the deviation of the isotopic ratio $^{13}C/^{12}C$ (sm) with respect to an internationally agreed upon standard material (st):

$$\delta^{13}C \equiv 1000 \cdot \frac{\left[\left|\frac{^{13}C}{^{12}C}\right|_{sm} - \left|\frac{^{13}C}{^{12}C}\right|_{st}\right]}{\left|\frac{^{13}C}{^{12}C}\right|_{st}} = 1000 \cdot \left[\frac{\left|\frac{^{13}C}{^{12}C}\right|_{sm}}{\left|\frac{^{13}C}{^{12}C}\right|_{st}} - 1\right]. \tag{3.3}$$

Typical $\delta^{13}C$ values vary between $+2‰$ to $-27‰$ depending on the fractionation history of the specific sample material. A negative value $\delta^{13}C$ means that the sample is isotopically lighter than the standard probe. A positive value means that the sample is enriched in the heavier isotope components with respect to the standard. The standard material is the fossil belemnite from the Pee Dee formation in South Carolina, PDB, $(^{13}C/^{12}C)_{PDB} = 0.0112372$.

For a material with the average $(^{13}C/^{12}C)_{sm}$ ratio of 0.01112235, the value of the fractionation parameter would be $\delta^{13}C = -10.221‰$. A reduction in the $^{13}C/^{12}C$ ratio by 1.8% would shift this value to $\delta^{13}C = -28.036915‰$. This is a rather dramatic change that needs to be accounted for.

For the oxygen isotopes, the international standard is the isotopic ratios of ^{18}O or ^{17}O to ^{16}O in ocean water (SMOW). [$^{18}O/^{16}O = 2005.20 \pm 0.43‰$ (a ratio of 1 part per approximately 498.7 parts) and $^{17}O/^{16}O = 379.9 \pm 1.6‰$ (a ratio of 1 part per approximately 2632 parts)]:

$$\delta^{18}O = \left(\frac{\left(\frac{O^{18}}{O^{16}}\right)_{sample}}{\left(\frac{O^{18}}{O^{16}}\right)_{SMOW}} - 1\right) \cdot 1000‰. \tag{3.4}$$

For higher ^{18}O values compared to the SMOV standard, the fractionation value is positive, $\delta^{18}O > 0$ and for lower ^{18}O values with respect to the standard, the fractionation value is negative, $\delta^{18}O < 0$.

For analyzing other materials, such as lead standard, probes have been produced and managed by the National Institute of Standards (NIST). For lead, the isotopic standards are SRM Pb 981, 982, and 983, but more standards are being developed to improve the sensitivity of the approach for different material compositions (https://www.nist.gov/publications/certification-report-reference-materials-erm-ae142-and-erm-eb400-pure-pb-solution-and). These international standards are of crucial importance, allowing for various fractionation measurements to be normalized and reliably compared.

3.4 EXAMPLES OF ISOTOPE FRACTIONATION IN ANTHROPOLOGY

In the following section, we will discuss a few examples that demonstrate the power of isotope fractionation measurements for exploring the living conditions and trading habits of people past. Since there are a multitude of isotopes associated with the chemistry of human life and habits, a multitude of possibilities exist for mapping them by following the specific isotope chemistry and its reflections in human remains. We will select only a few of these samples from a vast range of the scientific literature addressing this topic.

3.4.1 ANCIENT EATING HABITS

As suggested by the ^{13}C isoscape map, shown in Figure 3.8, there are two different processes for photochemical assimilation of CO_2 in plants (photosynthesis cycles), which are based on different chemistry patterns. This leads to different carbon fractionation values $\delta^{13}C$ in plants and the associated food chains, peaking at values of $\delta^{13}C = -26.5‰(C_3)$ and $\delta^{13}C = -12.5‰(C_4)$, respectively, as shown in Figure 3.11.

While C_3 plants with lower $\delta^{13}C$ values dominate the northern, cooler regions of Europe and North America, the habitat of C_4 plants with higher $\delta^{13}C$ values is predominant in the warmer regions of South and Central America, Africa, and Australia (Figure 3.9a). These ratios are subsequently reflected in the $\delta^{13}C$ values of herbivore animals in the respective regions, as demonstrated in Figure 3.9b. A mixture of C_3 and C_4 diet would translate into intermediary fractionation values. These values are then also transferred to carnivores' since they primarily consume local animals.

Plants on the North American continent are predominantly C_3 plants, translating into fractionation values of $\delta^{13}C = -21.4‰$ in the bone collagen of herbivores and also carnivores and humans. This can be followed in the example of maize, which is a typical C_4 plant that migrated north from the Mesoamerican regions. A 10% addition of maize to an otherwise C_3 dominated diet would increase the $\delta^{13}C$ value in the human body material to $\delta^{13}C = -20\%$. A diet based on maize only would result in pure C_4 values of $\delta^{13}C = -10‰$. This is beautifully demonstrated in the analysis of the carbon isotope ratios of indigenous tribes in North America. The values result from the bone analysis of human skeletons, where the increase in $\delta^{13}C$ from

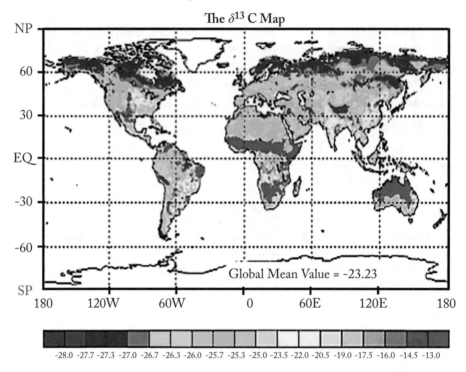

Figure 3.8: The global ^{13}C fractionation map (isoscape) showing vast differences in the ^{13}C/^{12}C isotope distribution ratios ranging from $\delta^{13}C = -28‰$ to $-13‰$. These variations are the result of climate conditions and different levels of vegetation https://agupubs.onlinelibrary.wiley.com/doi/full/10.1029/2003GB002141.

$-20‰$ to $-10‰$ indicates just such a transition in food habits over the period from 500 BC to 500 AD. During this period, the tribes gradually converted to the consumption of corn with the migration of that plant north, as shown in Figure 3.10. At 1500 AD, the nutrition was to 75% based on corn consumption [10].

Similar phenomena can also be observed with the change to seafood consumption, since ocean-based food chains are characterized by different fractionation processes, leading to a fractionation value of $\delta^{13} = -18‰$, which is substantially different from a value associated with a pure C$_3$- or C$_4$-based diet (Figure 3.11). Differences in δ^{13}C fractionation values can also be observed in the analysis of the skeleton material of early indigenous populations with different food habits. The analysis of bone material of people from coastal areas (British Columbia) shows a value of $\delta^{13}C = -13.4 \pm 0.9‰$, which suggests a nearly 100% seafood-based nutrition. In comparison, skeletal material from indigenous tribes in central British Columbia yields fractionation values of $\delta^{13}C = -15.4 \pm 0.3‰$, pointing to a mixed diet of about 65% seafood

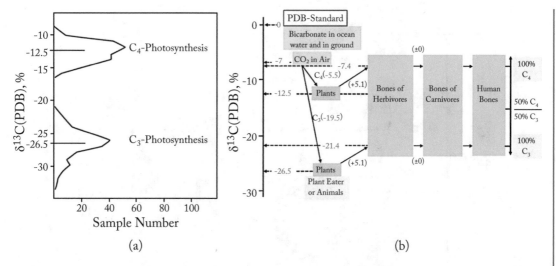

Figure 3.9: $\delta^{13}C$ distribution for C_3 plants in the colder regions of the northern and southern hemispheres and C_4 plants in warmer tropical regions (a) and the processing of food and the conversion to measurable $\delta^{13}C$ values in human remains (b). Figure plotted based on the data presented in N. Van der Merve, *American Scientist*, 70:596, 1982.

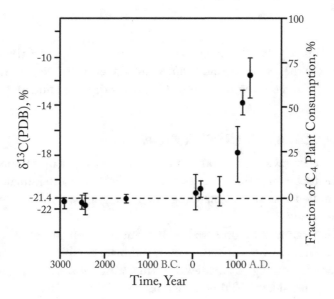

Figure 3.10: Change of the $\delta^{13}C$ fractionation value in the bone material of indigenous people in woodland North America, indicating a substantial change of diet from C_3- to C_4-based food intake. Figure plotted based on the data presented in N. Van der Merve, *American Scientist*, 70:596, 1982.

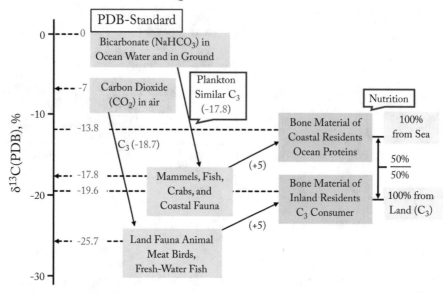

Figure 3.11: The $\delta^{13}C$ distribution in sea fauna and flora and its conversion to measurable $\delta^{13}C$ values in human and animal remains. Figure plotted based on the data presented in N. Van der Merve, *American Scientist*, 70:596, 1982.

(salmon) and approximately 35% C_3 plant-based nutrition. Analysis of the skeletons of early population groups from the region around Ottawa (inland residents) yields a fractionation value of $\delta^{13}C = -19.6 \pm 0.9‰$, pointing to $\sim$ 100% C_3 plant-originated nutrition.

3.4.2   HUMAN MIGRATION PATTERNS

Since environmental changes such as food or climate cause changes in isotope fractionation, its analysis is a powerful tool in mapping changes in human habit and also human migration. In the following we present two examples of mapping human migration based on carbon and oxygen isotope fractionation analysis.

This correlation can also be observed in the fractionation values of the skeletal material of Norsemen or Vikings who, depending on the climate conditions, changed from seafood intake to an agrarian diet, during the medieval warm period from 1000 AD to 1200 AD (Figure 3.12c). During this period, the Vikings converted their lifestyle after centuries of seafaring to the establishment of farming settlements (Figures 3.12a and b). With the onset of the Little Ice Age after 1200 AD, agrarian productivity declined, forcing people to switch back to a seafood dominated diet, as suggested by the $\delta^{13}C$ fractionation values displayed as a function of age in Figure 3.12d [11]. The corresponding age was determined by the carbon-14 method described

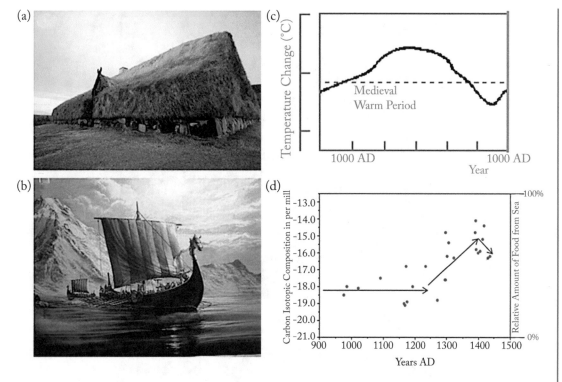

Figure 3.12: Viking longhouse on Iceland (a) and long ships (b) in coastal waters. Temperature change in Europe between 900 and 1500 AD, indicating the medieval warm period and the transition to the Little Ice Age (c). This correlates with the change of the $\delta^{13}C$ fractionation values from $-18‰$ to $-15‰$, reflecting a change from primarily seafood to agrarian products among people from coastal regions (d) (http://www.ancientpages.com/2016/04/16/viking-longships-fearless-dragonships-daringthe-oceans-and-seas/).

in Chapter 5. Detailed studies on Viking samples have been performed to trace the living and migration patterns of these early seafaring Norseman tribe [12].

Besides investigation of the carbon isotope abundance distribution in human remains, the abundance ratios of other isotopes may also provide important information, demonstrated in the two subsequent examples. The most important is the fractionation of oxygen isotopes, which occurs with rainfall as a result of miniscule differences in mass and therefore gravity during rainfall. In general, the ^{18}O fraction decreases with altitude and distance from a coast, since heavy ^{18}O water condenses and freezes out faster than ^{16}O-based water. Therefore, oxygen isotope fractionation has emerged as a powerful tool in climate science and plays an important role in anthropology, pointing to the origins of migrating groups.

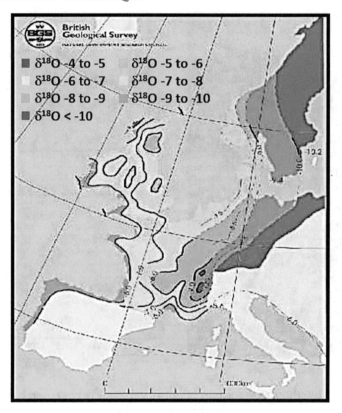

Figure 3.13: The $\delta^{18}O$ fractionation pattern for Northern Europe as established by the ^{18}O composition of groundwater and rainwater https://www.caitlingreen.org/2015/10/oxygen-isotope-evid.

A well-known example is the case of the Stonehenge Archer, the 2002 discovery of a gravesite near Stonehenge that contained two well-conserved skeletons with an assortment of rich burial goods, ranging from jewelry to weapons, near Stonehenge. The age of the bones was determined by radiocarbon dating to be from 2300 BC. While this was initially interpreted as the existence of a ruler of the Stonehenge site, oxygen abundance analysis painted a different picture. Due to their specific climate conditions, the British Isles have a peculiar oxygen fractionation pattern. The rain in coastal regions contains a higher ^{18}O abundance than rain in continental regions. The fractionation value from the west coast of England to the east coast varies between $\delta^{18}O \approx -4$ to $-8‰$.

Tooth enamel in human remains typically maintains the isotope ratio formed at the childhood location. If the persons found at the gravesite were from the British Isles, their tooth enamel would have reflected a similar high $\delta^{18}O$ fractionation value (Figure 3.13). However, study of the tooth enamel of the skeletons indicates a very low $^{18}O/^{16}O$ ratio: $\delta^{18}O \approx -9$ to -10, much lower than typical for the British Isles [13]. This result, in combination with the analysis of the strontium isotope composition in the tooth material, clearly indicates that the buried persons were not local but must have originated in northern alpine regions of central Europe, a clear indication of the close connection between the people on the Isles and those in Central Europe during the Bronze Age. Isotope analysis of the grave goods further supports this claim—the copper-based goods had originated in Spain, while the gold was from central Eastern Europe, most likely from the area of present-day Romania—suggesting a wide inter-European trade pattern at these early times. On the basis of isotope analysis of pig bones, Stonehenge itself during Neolithic times was identified as a cultural and possibly religious center, which attracted people from all over the British Isles for annual meetings and feasts at a scale unimagined before, adding new understanding on the role of this important monument [14].

3.4.3 THE APPLICATION OF HEAVIER ISOTOPES

While fractionation of light isotopes is due to environmental causes, the fractionation of isotopes of heavy elements such as strontium or lead is typically caused by geochemical or natural radioactivity induced processes. Such studies are often complementary to the analysis of light isotopes; the isotope distribution of strontium in the archer's remains confirmed the results of the oxygen isotope analysis.

Strontium behaves chemically like calcium and therefore can be found enriched in bone material, since Sr can replace Ca by food or water intake. This is also evident from the deposition of the long-lived ^{90}Sr ($T_{1/2} = 30$ years) that was emitted during the nuclear weapons test program, causing an overall slight enrichment in ^{90}Sr radioactivity of the baby boomer generation. Similar enrichments can be found because of the Chernobyl nuclear power plant accident.

The element strontium has four stable Sr isotopes: ^{84}Sr (0.56%), ^{86}Sr (9.86%), ^{87}Sr (6.94%), and the most abundant one ^{88}Sr (82.58%). The average ratio of the two stable isotopes ^{86}Sr and ^{87}Sr is $^{87}Sr/^{86}Sr = 0.704$. However, stable ^{87}Sr is enriched by decay of long-lived rubidium ^{87}Rb with a half-life of $T_{1/2} = 4.9 \cdot 10^{10}$ years, which is ten times longer than the age of earth. This rubidium isotope has been produced by neutron capture reactions in the helium and carbon burning of massive stars prior to the formation of the solar system and was implemented in earth material.

Rubidium behaves chemically like potassium in minerals and is therefore widespread. 87Rb has been used extensively in dating geological samples but, since plants easily absorb it as a potassium-like nutrient, the enrichment is transferred to biological systems; ^{87}Rb decays to stable ^{87}Sr by β-decay, with emission of an electron ejected from the nucleus. An environment with high Rb content therefore causes an increase in the $^{87}Sr/^{86}Sr$ ratio. The process of isotope enhancement through radioactive decay is called radiogenic infusion. The $^{87}Sr/^{86}Sr$ ratio in tooth-enamel materials is therefore related to the rubidium composition of geological material.

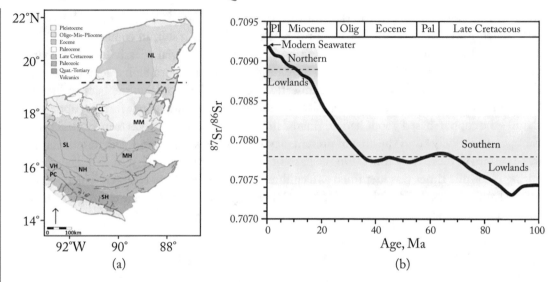

Figure 3.14: The map in (a) shows the subregions and the variance of the measured ^{87}Sr/^{86}Sr values. Northern Lowlands (NL), Southern Lowlands (SL), Volcanic Highlands (VH), Pacific Coast (PC), Metamorphic Province (MP), and Maya Mountains (MM); (b) shows the ^{87}Sr/^{86}Sr isotope ratio along a north-south axis of the Yucatán Peninsula as indicated on the map shown in (a). The pattern shows a gradual decline in ratio due to the geological formation of the area. Reprinted with permission from [15]. Copyright 2004 Elsevier.

The abundance of non-radioactive ^{86}Sr in minerals is constant; ^{87}Sr is stable but continuously produced by decay of the radioactive ^{87}Rb. While continental rock is rich in ^{87}Rb, limestone typically contains slightly less ^{87}Rb, which is reflected in the ^{87}Sr/^{86}Sr ratios. This is not only the case for the British Isles but also for the Yucatán Peninsula separating the Gulf of Mexico from the Caribbean Sea. The bedrock of the peninsula consists of limestone sediments with marine sediment inclusions, a variation that is reflected in the change of the local ^{87}Sr/^{86}Sr ratio.

The northern section of the Yucatán Peninsula was formed by Ca/Rb minerals containing marine sediments and shows an ^{87}Sr/^{86}Sr enrichment that is comparable to the ratio in seawater. However, this enrichment declines toward the south, as shown in Figure 3.14 [15].

The variations in ratio between the different regions along the Yucatan peninsula are shown in Figure 3.14 for specific regions: Northern Lowlands (NL), Southern Lowlands (SL), Volcanic Highlands (VH), Pacific Coast (PC), Metamorphic Province (MP), and Maya Mountains (MM). These variations can serve for the study of migration patterns of Mayan and pre-Mayan population groups.

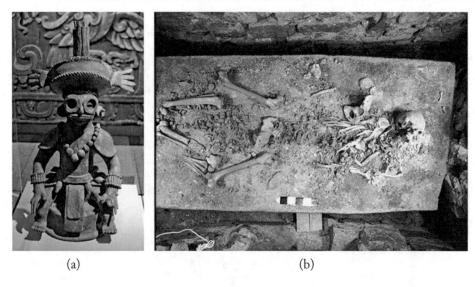

(a)   (b)

Figure 3.15: A figurine of Yax K'uk Mo with goggle features that are considered to be characteristic for the Teotihuacán culture (a) (https://www.ancient.eu/Yax_Kuk_Mo/). Photograph of the gravesite of Yax K'uk Mo (b) (https://www.cgu.edu/news/2018/10/david-sedat-maya-copan-cacao-chocolate/).

A prime example is the origin of Yax K'uk Mo, who ruled the City of Copan in southern Yucatán during the 5th century AD. After his coronation in 426 AD, he converted the town from a small village to a center of Mayan culture, within only decades. The origin of Yax K'uk Mo is unknown; his name is translated as "Radiant First Quetzal Macaw" or "Sun-Eyed Green Macaw" or as "Sun in the Mouth of the Quetzal Bird," indicating foreign origins. In view of his rapid ascendance, it was speculated that he was installed by the Teotihuacán culture in Central Mexico to extend its political influence south. This assumption was based on specific features of the statues of Yax K'uk Mo, showing him wearing goggles (Figure 3.15a) characteristic for Teotihuacán figurines [16].

The gravesite of Yax K'uk Mo was discovered and excavated at Copán in 2000 (Figure 3.15b). A detailed isotope analysis of the bones and tooth enamel were performed. The isotope analysis revealed that the individual spent his early years near Tikal in the Petén basin region ($^{87}Sr/^{86}Sr \approx 0.708$) and then moved to an area located between Tikal and Copán ($^{87}Sr/^{86}Sr \approx 0.707$). The $^{87}Sr/^{86}Sr$ isotope ratio does not match the much lower ratio of Teotihuacán in the volcanic highlands of Central Mexico, which is recorded as $^{87}Sr/^{86}Sr = 0.704$.

Lead was one of the earliest metals discovered by humans and was in use by 3000 BC in ancient Rome the use of lead is recorded for multiple applications and purposes. The Latin word for lead, "Plumbum," provides the basis for the chemical term Pb for this element. The

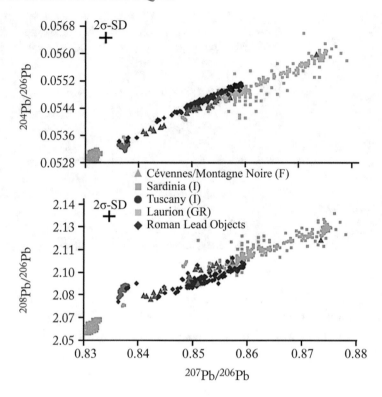

Figure 3.16: Pb isotope ratios for Roman lead objects from Augustan military camps in Germania are plotted in comparison to lead samples from ore-mining districts in Greece, France, and Italy. Reprinted with permission from [17]. Copyright 2009 Springer Nature.

importance of lead in the life of Rome cannot be overestimated. Lead use ranged from the production of water pipes and the lining of baths with this soft metal, in early times, to use for roofing, coffins, cisterns, tanks, and gutters in the Middle Ages. Lead was used in weaponry, coin minting, and in the preparation of pigments, paints, and cosmetics. Roman life was contaminated with lead, and many historians argue that Rome's decline was due to the general lead poisoning of its population (https://www.ila-lead.org/UserFiles/File/factbook/annex.pdf). Roman winemakers used lead pots or lead-lined copper kettles for crushing the grapes to enhance the sweetness of the wine. Lead was also considered an important sweetener for a large variety of other food items, as demonstrated by the collection of Roman recipes in the cookbook De re coquinaria (On the Subject of Cooking) presumably first collected by Marcus Gavius Apicius, the famous Roman gourmet who was described by Pliny as "the most gluttonous gorger of all spendthrifts" (https://www.knowtheromans.co.uk/Sources/Apicius/).

This extensive need for lead was the basis for an extensive lead trading business that flourished throughout antiquity. Lead mines were located all over the Roman Empire, from which ships loaded with lead ingots distributed these mining and smelting products throughout the Mediterranean. Many of the ships were lost at sea and their loads provide a wealth of information for the historian of antique trading routes and patterns.

These findings are useful not only in the analysis of lead ware but also in the analysis of the spurious lead content in other metal ingots, such as brass and tin.

Lead findings also provided Roman military information as shown in Figure 3.16, which demonstrates the lead isotope characteristics of Roman objects found at Augustan military camps in Germania. The comparison of the isotope ratios with the isotope ratios of known lead mines in the Cevennes (today's France), and in Tuscany and Sardinia (today's Italy), as well as with lead from Laurion (today in Greece), suggests that the Roman troops, stationed in Germania, were mostly recruited from the western part of the empire [17].

3.5 REFERENCES

[1] N. Reifarth, Archäometrische untersuchungen am münzschatz von bell, qualitative und quantitative elementanalyse mittels neutronenaktivierung am forschungszentrum karlsruhe, *Fachhochschule Erfurt FB Konservierung und Restaurierung Altonaerstr*, Sommersemester, 2002. 140

[2] J. Yellin, M. Broshi, and H. Eshel, Pottery of Qumran and Ein Ghuweir: The first chemical exploration of provenience, *Bull. Amer. Schools Orien. Res.*, 321:65–78, 2001. DOI: 10.2307/1357658. 141, 142

[3] A. Denker, C. Laurenze-Landsberg, K. Kleinert, and B. Schröder-Ameibidt, Paintings reveal their secrets: Neutron autoradiography allows the visualization of hidden layers, *Neutron Meth. Archaeol. and Cult. Herit.*, pages 41–52, Eds., N. Kardjilov and G. Festa, Springer, 2017. DOI: 10.1007/978-3-319-33163-8_3. 142

[4] M. W. Ainsworth, E. Haverkamp-Begemann, J. Brealey, and P. Meyers, with contributions by K. Groen, M. J. Cotter, L. van Zelst, and E. V. Sayre, Art and autoradiography: Insights into the genesis of paintings by Rembrandt, Van Dyck, and Vermeer, *Metropol. Mus. Art*, pages 12–18, New York, 1987. DOI: 10.2307/1506019. 144

[5] S. Nystad, Der Goldhelm, Jahrbuch der Berliner Museen, 41. Bd., (1999), pages 245–250 and Kristin Bahre u. a. (Hrsg.): *Rembrandt. Genie auf der Suche*. DuMont Literatur und Kunst, Köln 2006. 146

[6] E. Andersand and N. Grevesse, Abundances of the elements: Meteoric and solar, *Geochim. Cosmochim. Acta*, 53:197–214, 1989. 147

[7] M. Lugaro, U. Ott, and A. Kereszturi, Radioactive nuclei from cosmochronology to habitability, *Prog. Part. Nucl. Phys.*, 102:1–47, 2018. DOI: 10.1016/j.ppnp.2018.05.002. 147

[8] G. D. Farquhar, I. J. R. Ehleringer, and K. T. Hubick, Carbon isotope discrimination and photosynthesis, *Annu. Rev. Plant Physiol. Plant Mol. Biol.*, 40:503–537, 1989. DOI: 10.1146/annurev.pp.40.060189.002443. 147

[9] A. P. Fitzpatrick, The Amesbury archer, *Curr. Archaeol.*, 184:146–152, 2003. DOI: 10.1017/s0003598x00090980. 147

[10] N. Van den Merve and J. Vogel, ^{13}C Content of human collagen as a measure of prehistoric diet in woodland North America, *Nature*, 276:815–816, 1978. DOI: 10.1038/276815a0. 152

[11] J. Arneborg, J. Heinemeier, N. Lynnerup, H. L. Nielsen, N. Rude, and A. E. Sveinbjornsdottir, Change of diet of the Greenland Vikings determined from stable carbon isotope analysis and ^{14}C dating of their bones, *Radiocarbon*, 41:157–168, 1999. DOI: 10.1017/s0033822200019512. 154

[12] J. Montgomery, V. Grimes, J. Buckberry, J. A. Evans, M. P. Richards, and J. H. Barrett, Finding Vikings with isotope analysis: The view from wet and windy Islands, *J. North Atlantic*, 7:54–70, 2014. DOI: 10.3721/037.002.sp705. 155

[13] J. A. Evans, C. A. Chenery, and A. P. Fitzpatrick, Bronze age childhood migration of individuals near Stonehenge, revealed by strontium and oxygen isotope tooth enamel analysis, *Archaeometry*, 48:309–321, 2006. DOI: 10.1111/j.1475-4754.2006.00258.x. 157

[14] R. Madgwick, A. L. Lamb, H. Sloane, A. J. Nederbragt, U. Albarella, M. Parker Pearson, and J. A. Evans, Multi-isotope analysis reveals that feasts in the Stonehenge environs and across Wessex drew people and animals from throughout Britain, *Sci. Adv.*, 5:1–12, 2019. DOI: 10.1126/sciadv.aau6078. 157

[15] D. A. Hodell, R. L Quinn, M. Brenner, and G. Kamenov, Spatial variation of strontium isotopes (^{87}Sr/^{86}Sr) in the Maya region: A tool for tracking ancient human migration, *J. Arch. Sci.*, 31:585–601, 2004. DOI: 10.1016/j.jas.2003.10.009. 158

[16] C. Day, Isotopic analysis of teeth and bones solves a mesoamerican mystery, *Phys. Today*, 57:20–21, 2004. DOI: 10.1063/1.1650060. 159

[17] M. Bode, A. Hauptmann, and K. Mezger, Tracing Roman lead sources using lead isotope analyses in conjunction with archaeological and epigraphic evidence—a case study from Augustan/Tiberian Germania, *Archaeol. Anthropol. Sci.*, 1:177–194, 2009. DOI: 10.1007/s12520-009-0017-0. 160, 161

CHAPTER 4

Imaging Techniques

4.1 INTRODUCTION

The Roentgen Rays, The Roentgen Rays What is this craze, The town's ablaze, With the new phase of X-rays ways I'm full of daze, Shock and amaze, For nowadays, I hear they'll gaze, Thro' cloak and gown—and even stays, These naughty, naughty Roentgen Rays.

(Charles Kingsley, Photography, 1896)

Imaging techniques have always been of the utmost importance in investigating cultural heritage objects and artifacts. These techniques allow for the collection, analysis, and visualization of objects in their different parts, from macro- to atomic-scale levels. The simplest and most popular imaging method is photography, which has been used for more than 150 years. It enables visual depiction of objects on film or in digital formats. Photography is still the most used method in cataloging, increasing public access, education, and marketing of cultural heritage objects. The scale of photographic imaging at the macroscopic level can be divided into three main categories: *aerial photography, diagnostics of architectural sites, and single-object imaging*. Aerial photography is a broad imaging survey used to map the locations of different objects in archaeological excavations. Architectural imaging can be used for the structural mapping of buildings and thermography, for conservation purposes. Single-object imaging deals with the visualization of the shape and the integrity of a particular object, for cataloging and inventory purposes.

Another aspect of imaging is to reveal the smaller parts of an object for characterization purposes. This type of imaging can be performed through the mesoscopic, microscopic, or atomic levels to aid in conservation efforts and the interpretation of objects. Modern imaging techniques can be classified by their ability to probe the surface or internal structure of the object (or material). Methods for revealing internal structure are based on the penetrating of electrons, X-rays, or neutrons through the object (or sample). Some imaging techniques, such as reflectometry and radiography, do not require sample preparation. However, the microscopy methods do require sampling and sample preparation. In addition, we have already described several chemical mapping-based imaging methods (XRF, PIXE, EDS) in Chapter 2. These methods allow visualization of the distribution of selected chemical elements on the surface, using X-rays. Other methods exploit the reflections of UV, Visible, and IR light to extract information on the surface of different painted objects (reflectography). This chapter concentrates on three major groups of methods that are widely used in cultural heritage and artifact analysis: reflectography, microscopy, radiography, and tomography methods.

4.2   REFLECTOGRAPHY

4.2.1   GENERAL CONSIDERATIONS

This method is suited for the investigation of paintings. The significant components of all paintings are preparatory ground layers, sketch, multiple painting layers, and varnish. Most basic sketches were conducted using a black/brown pigment (charcoal, graphite, black bone, hematite). Frescoes and wall paintings have similar layouts; the support is the wall, while the color layers are applied to the sketched preparatory layer. Different parts of the paintings often undergo different evolutionary processes and modifications over time. Reflectography methods employ reflected light of different wavelengths to investigate the surface structure and modifications of the paintings.

UV radiation originates luminescence from the surface components, mainly organic polymeric varnishes and pigment binders. The luminescence of the light depends on the polymerization degree. This identification allows for estimating the age of the varnish, detecting surface modification, and identifying the repaired areas. Visible light images allow visualizing of the surface-pigment color distributions. Shining visible light in grazing incidence can help to characterize imperfections of the surface layers. IR light is the most used method in reflectography. The gypsum or calcite ground layers in paintings backscatter the IR light, while graphitic carbon or similar material, when used to sketch on the ground level, strongly absorbs light. The transparency of some pigments to IR radiation allows for identifying hidden details that have been painted with IR light-absorbing pigments; penetration depends on the IR radiation wavelength and the thickness of the paint layers.

The instrumentation of IR reflectography contains sources of infrared radiation and an infrared camera, which captures the light that reflects from the painting's surface. Software is used to processes the captured digitized image. The cameras contain CCD, indium gallium-antimonide (InGaAs), or platinum silicide (PtSi) detectors. CCD cameras have a smaller spectral sensitivity and cover both the visible light and the near-IR spectrum (below one micrometer). These cameras have a limited ability to explore deeper layers of paintings. Cameras with InGaAs and PtSi detectors have more extensive spectral sensitivity ranges, 0.9–1.7 μm and 1.2–5 μm, respectively, and capture light reflected from deeper layers.

4.2.2   EXAMPLES OF REFLECTOGRAPHY

An example of IR reflectography is found in the imaging of a painting on a wooden board (15th century) depicting St. Antonio Abbot (Figure 4.1a). This painting is part of a polyptych from the Church of S. Caterina Alessandria in Galatina (Lecce, Italy) [1]. An IR camera was used to acquire images in the visible spectrum RGB color (400–700 nm), and PAN modes, as well

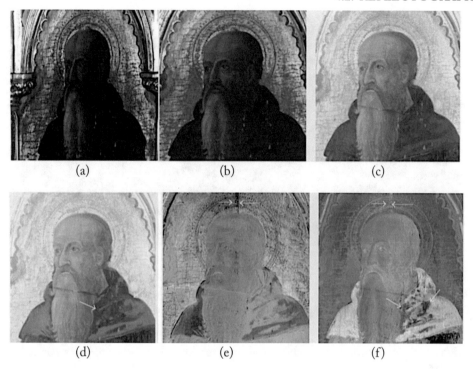

(a)   (b)   (c)

(d)   (e)   (f)

Figure 4.1: Input RGB (a) and PAN (b) visible images, IR1 (c) and IR2 (d) images and digitally processed PC (e), and NDI (f) images for wooden boar painting of St. Antonio Abbot from the Church of S. Caterina Alessandria in Galatina (Lecce, Italy) https://www.hindawi.com/journals/ijge/2011/738279/.

as in the near IR spectral bands at 700–950 nm (IR1) and 950–1150 nm (IR2). Two 60 W halogen lamps were used to illuminate the painting surface at 45°. To overcome instrument limitations, a post-capture principal component (PC) analysis was used. A linear combination of two spectral bands (IR1 and RI2) was also conducted to obtain quantitative measures of the surface properties. The normalized difference index (NDI) is computed by the NDI = (IR2 − IR1)/(IR2 + IR1) formula. The results of imaging and post-processing are shown in Figure 4.1.

The PAN and RGB images highlight the painting's conservation state, with cracks and superficial deposits (Figures 4.1a and b). The infrared images (Figures 4.1c and d), in particular the IR2 channel, show new features (darker grey) on the Saint's tunic and mantle that are referable to later painting layers, with the classical effect of so-called *leopard spots*. The PC image (Figure 4.1e) shows additional information on the mantle and the tunic, while some features on the face have disappeared. This image also shows more features of degradation in the wooden board on top of the saint's halo. The application of the NDI (Figure 4.1f) confirms the features

(a)   (b)   (c)

Figure 4.2: Visible (a), IR (b), and IRFC (c) images from the God's face in the painting God the Father with Angels, by Giotto. Reprinted with permission from [2]. Copyright 2019 Elsevier.

in the right part of the mantle. Such a nonhomogeneous dark area suggests the existence of restoration work.

IR false color (IRFC) imaging is a powerful approach for combining the green and red channels of the visible RGB colors with the infrared image. This photographic technique allows for the identification of pigments with different compositions that have similar colors. An example of such imaging is the FCIR investigation of *God the Father with Angels* painted by Giotto Di Bondone in approximately 1330. This masterpiece is a tempera and was part of the altarpiece of the Santa Croce church (Florence), now held in the San Diego Museum of Art [2]. XRF analysis shows that the blue hat of God the Father is painted with ultramarine (Figure 4.2a). This pigment became almost transparent in the near IR image (Figure 4.2b), except for a few darker areas from restoration on the right-hand side. These features and other areas where some material was lost in IR imaging, become well highlighted in the IRFC image (Figure 4.2c). The IR reflectography image clearly shows the original sketch by Giotto, and also contains details (round dark spots) on the hat, which had later been decorated with gold. The IR image shows the typical drawing techniques of Giotto with thin and large brushstrokes made with carbon-based pigment.

4.3 RADIOGRAPHY AND TOMOGRAPHY

4.3.1 GENERAL CONSIDERATIONS

The discovery of X-ray by Wilhelm Röntgen in 1895 originated a new diagnostic technique known as radiography. Newly discovered X-rays and radiography were immediately used in medical diagnostics. X-ray radiography was also quickly put to use in archaeology for the investigation of different objects.

Advanced radiography is a simple method and allows for the visualization of the density distribution within an object by differential absorption of X-rays (γ-rays or neutrons). This can be performed by recording the X-rays that are passing through an object on a photographic film, area detector, fluorescent screen, or CCD camera. The instruments vary depending on the radiation source, resolution, size of the detector, and the dimensions of the studied object. The most common instrumentation contains a point X-ray source with a conical beam geometry. Other specific techniques involve the use of synchrotron X-ray radiation or high energy industrial X-ray sources. Radiography offers the possibility of obtaining insight into the internal structure of objects.

X-ray radiography is a relatively affordable method and allows for answering multiple questions about preparation, function, and the state of preservation of different objects. This technique sometimes provides information that is not possible to obtain by any other method. In addition to X-rays, neutrons and γ-rays can also be employed to image the internal structure of cultural heritage objects and historical artifacts. The use of neutrons in radiography was first considered in 1935, just three years after their discovery. The practical application of neutron radiography is rare because neutrons are less available, as compared with X-rays. The cobalt-60 isotope is a common source for producing high-energy (1.17 and 1.33 MeV) γ-rays, which can be used in radiography. γ-ray sources can be portable and operated without electricity or a water-cooling system. These sources continuously emit γ-rays but, over time, gradually lose activity. For safety, these sources must be contained in containers that are shielded with heavy metals. However, the source should be removed from the container during radiography. γ-rays have seldom been used for art objects and artifact radiography.

X-rays (and γ-rays) may be transmitted through different materials without a change of direction or a significant loss of energy. To perform successful radiography imaging, the X-rays should be attenuated when passing through the object. The attenuation degree depends on the energy of the X-rays, and the density and thickness of the materials. The attenuation of X-rays encompasses absorption and scattering processes. The attenuation of an X-ray beam in a material is exponential. It can be described by the following relationship (also see Chapter 2):

$$I_x = I_0 e^{-\mu x}, \tag{4.1}$$

where I_0 and I_x are intensities at incident X-ray beam and at the x depth, respectively. The e term is the natural logarithm base, and μ is the linear attenuation coefficient. This relationship indicates that a given thickness of the material will attenuate a fixed proportion of the incom-

Table 4.1: Half-thickness in millimeters for materials with different density (ρ, g/cm^3) and X-ray energies (keV)

X-Ray Energy (keV)	H$_2$O (ρ=1)	Al(ρ=2.7)	Cu(ρ=8.9)	Lead(ρ=11.2)
102	1160	20	3.9	0.4
154	530	16	1.7	0.1

ing X-ray beam. Therefore, a half-thickness concept can be introduced, which is defined as the thickness of a material that reduces the incident X-ray intensity to half of its initial intensity. Table 4.1 illustrates that different materials attenuate X-rays differently. For example, lead heavily absorbs X-rays due to its high density (atomic number). The materials containing lighter elements attenuate X-ray less strongly. Low-energy (softer) X-rays are attenuated more strongly than hard X-rays.

Optimization of radiography requires the selection of the X-ray beam energy appropriate to the thickness of the object. Another parameter that should be optimized is the magnification (M) factor:

$$M = \frac{d_{ss} + d_{sd}}{d_{ss}}, \tag{4.2}$$

where d_{ss} and d_{sd} are the distances between the source to the sample and between the sample to detector, respectively. The greater the d_{sd}, the higher the resolution of the image. To obtain a high-resolution image, one should use an intense X-ray source and long data collection times.

X-ray radiography has limitations when imaging massive and denser objects. For example, imaging of full-size marble and bronze statues is challenging. In such situations, the use of neutron radiography is a better choice. Neutrons are attenuated by materials that contain light elements (H, B, Li). X-ray and neutron radiography can be used as two complementary methods. A specific form of neutron radiography is based on neutron activation analysis, described in Chapter 3. This method, sometimes also called autoradiography, can be applied for analyzing pigments in painting. The nuclei in the artwork materials capture neutrons, producing multiple short-lived radioactive isotopes with different decay half-lives. The γ-rays emitted during the decays are recorded at different times. Such neutron-based imaging enables revealing paint layers applied on top of each other during the preparation of the artwork.

The mathematical treatment of a set of two-dimensional (2D) radiographic images allows the creation of a three-dimensional (3D) distribution model. This method is known as computed tomography (CT), which allows reconstruction and visualization of the object's internal structure. CT with X-rays, γ-rays, or neutrons is a convenient and valuable technique. The sets of 2D images are obtained in devices with transmission geometry. The resolution of the final reconstructed image depends on the source type, magnification factor, and size and resolution of the detector. The point X-ray source with a cone geometry and an area detector is the most typical laboratory CT instrument. Tomographic examination in materials science is performed by rotating the object around one axis at a fixed position between X-ray source and detector. The availably of a vast number of 2D images ensures high quality absorption contrast of the CT model.

The size of the object and the final resolution of the volume pixels (voxels) are connected in tomographic imaging. Meter-sized objects can be imaged with a resolution of about 1 mm. In contrast, millimeter-sized objects can be imaged with resolution below a micrometer. In synchrotron X-ray radiation-based tomography, nanometer-scale resolution can be obtained. Furthermore, the visualization of the internal structure can be considerably improved by using phase-contrast imaging. This technique, better known as holotomography, is based on the inhomogeneous refractive index and the thickness distributions of the object. This method also provides quantitative phase mapping, in addition to tomographic reconstruction. The working principle of neutron tomography is similar to X-ray tomography. The spatial resolution for this method is around 100 μm. The outstanding penetration ability of neutrons makes this method very useful for imaging dense and thick objects. The high sensitivity for light elements also allows the differentiation of water and organics in inorganic materials.

4.3.2   EXAMPLES OF RADIOGRAPHY AND TOMOGRAPHY

An Egyptian child mummy (Figure 4.3a) from the Senckenberg Museum of Natural History (Frankfurt am Main, Germany) was the first mummy investigated with X-rays. Recently, this mummy was reinvestigated using novel dual-source CT and radiocarbon dating methods [3]. Figure 4.3b depicts the historical X-ray radiogram of the mummy's knees, imaged in 1896. Later investigation compared the oldest and the newest radiographic technology, which was able to assess paleopathological and demographic information about the mummy. Old-style X-ray radiography has limited contrast for soft tissue, as evidenced in the 1896 image (Figure 4.3b) that was recorded on photographic film. The sides of the objects were interchanged during the negative-into-positive photographic transformation. Also, due to the not-proper positioning of the mummy between the X-ray source and photographic plate, different parts of the image have abnormal size ratios. Modern CT-based reconstruction of the mummy (Figure 4.3c) shows the correct anatomical position of the body parts. The advantage of CT compared to conventional X-ray radiography is that CT is capable of differentiating soft tissue along with internal organs, as well as allowing for the identification of embalming materials and wrappings. For example, many internal body parts, muscular tissues, organs, or their remains (nose, ossicles, auricles, bulb/lens, teeth, optic nerve, eye muscles, tentorium long biceps, tendon, and others) can be easily identified (Figure 4.3d). Brain fragments, spinal cord, and peripheral nerves can also be seen. Tomographic reconstruction (Figure 4.3e) demonstrates the realistic 3D image of the mummy with brain remnants inside the posterior cranial fossa and within the cervical spinal canal.

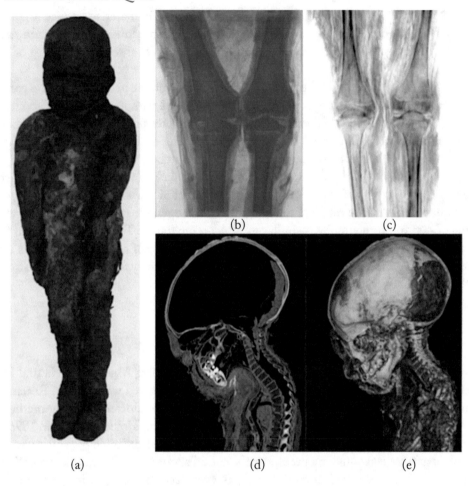

(a)   (b)   (c)

(d)   (e)

Figure 4.3: Egyptian child mummy (a) from the Senckenberg Museum of Natural History in Frankfurt. The historical X-ray image (b) of the mummy's knees. Modern CT-based reconstruction of the lower extremities (c). Sagittal multi-planar reformation (d), and volume reconstruction (e) demonstrate shrunken brain remnants of the child mummy. Reprinted with permission from [3]. Copyright 2016 Elsevier.

Dental markers were used for the chronological age estimation. Deciduous (milk) teeth had erupted and were well mineralized. In contrast, most of the molar teeth at the back of the mouth had not erupted. This fact indicates that the age of the mummy, at its time of death, can vary from between four and five years old. Some other skeletal markers suggest an even younger age (2–3 years old). However, dental development correlates more closely with age than the other markers. Therefore, the authors assumed that the person was four to five years old. The

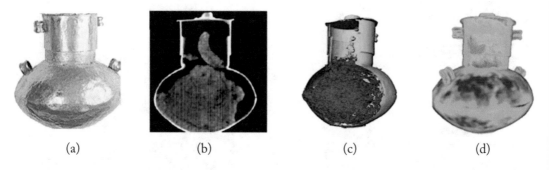

(a)   (b)   (c)   (d)

Figure 4.4: A golden vial (a), a 2D neutron radiography image (b), 3D reconstruction model (c), and color schematics for the density of the neutron attenuation degree (d) https://www.mdpi.com/2313-433X/4/2/25.

CT data identified the mummy as male. Radiocarbon dating gives the age of the mummy to be 2247 ± 23 years (a calibrated age interval, between 378 and 235 years BC). These results highlight the incredible progress from early X-ray radiology to CT-based X-ray imaging. This high-resolution imaging allows identification of body cavities, bones and their structures, soft tissues, age at death, and sex, along with wrapping and mummification techniques.

Neutron imaging is a complementary method to X-ray-based radiography and tomography. X-rays interact with the electron shell of target atoms. In X-ray-based methods, the increase of the atomic number in objects increases the attenuation coefficient. However, in neutron imaging, no clear correlation can be found with the atomic number. At the same time, heavy metals, for example lead, are practically transparent. Neutron imaging with 2D projections can be performed in a way similar to X-ray radiography. For in-depth information revealing the internal structure of the object, the CT approach can also be used. For example, a closed golden vial (Figure 4.4a) excavated from the "Chebotarev-5" archaeological site near Rostov-on-Don (Russia) was investigated by neutron CT to reveal its internal structure [4]. Tomography imaging was conducted on a pulsed (350 μs) neutron source beamline attached to a nuclear reactor. A beam of the thermal neutrons with wavelengths ranging from 0.02–0.8 nm and a spectral distribution maximum of 0.18 nm was used. The neutron flux at the object was $\approx 5.5 \cdot 10^6$ n/cm^2/s.

A set of neutron 2D projection images were collected, using a CCD detector with the 20 × 20 cm maximum field of view. A single neutron radiographic image of the vial shows a grey internal region that corresponds to a fossil filling the inside, as well as light areas of the golden walls (Figure 4.4b). Tomography reconstruction of the multiple image projections yielded a 3D model of the vial (Figure 4.4c). The fossilized remnants inside the jar are formless with a volume of 3.162 cm^3. In contrast, the overall volume of the gold material is 0.615 cm^3. A 3D model of the spatial distribution of the thickness deviation of the gold was also computed (Figure 4.4d). This model shows that deviations in the thickness of the golden wall are minimal.

(a)   (b)   (c)   (d)

Figure 4.5: A Buddhist bronze sculpture from the collection of the Rietberg Museum in Zurich (a), a 2D neutron radiography image (b), 3D reconstruction model (c), and a color model for the objects placed inside during preparation of sculpture (d).

Similar neutron imaging research was conducted for the examination of a Buddhist bronze sculpture (Figure 4.5a), from the collection of the Rietberg Museum in Zurich, with devotional objects placed inside during preparation [5]. Neutron imaging is the only option for such objects, as X-rays are unable to provide good contrast due to their low transmission. A single frontal neutron image of the sculpture clearly shows several objects placed within the sculpture (Figure 4.5b). The reconstructed model allows for a description of the objects (Figure 4.5c). Segmentation of this model enables virtual reconstruction and a separate investigation of the objects inserted in the statue (Figure 4.5d). Such detailed analysis pointed out three distinct forms; a *scroll* tied up with a cord and probably containing a religious text, a *heart-shaped object* inserted in the chest area, and a *pouch*. The heart shaped object is a small capsule wrapped in a piece of cloth and filled, in part, with a less attenuating material. The pouch consists of a couple of smaller spherical objects. Such neutron CT imaging provides information that cannot be obtained with any other method.

4.4   MICROSCOPIC METHODS

4.4.1   GENERAL CONSIDERATIONS

Microscopy involves the study of objects that are too small to be examined by the human eye. Light microscopes were first developed in the early 17th century. Modern compound optical microscopes (OM) contain a minimum of two lenses: an objective and an eyepiece. The objective is placed close to the object under study while the eyepiece is placed relatively close to the eye. The magnification of the light microscope can be increased by employing multiple lenses.

The resolution of an optical microscope is limited by the diffraction of light and the optical aberrations of the lenses.

Fine grinding of the lens' surface allows correcting aberrations. An option to improve the resolution is to use an oily objective lens. A drop of a transparent oil is placed between the sample and objective, which allows a $\sim$ 35% improvement in resolution. A more noticeable improvement in resolution could be achieved by using a UV light with a 100–300 nm wavelength. The regular glass lens absorbs most of the UV light. Therefore, the focusing lenses should be made, in this case, of UV transparent materials such as quartz or lithium fluoride.

Light microscopes can be in two forms: biological and metallographic. The biological microscope requires an optically transparent sample. Using a lens, the light from a lamp is focused through the sample and the microscope creates a real image on the attached camera. The metallurgical microscope is used for examining many cultural heritage objects that cannot easily be made thin enough to be optically transparent.

For most compound OMs, the maximum magnification can be $\sim$ 1,500 times. Beyond this magnification, objects become fuzzy. The electron microscopy allows high (x 1,500–100,000) and ultra-high (above 100,000 times) magnification of the surface of cultural heritage objects. An advantage of these methods is a considerable depth of field that allows rough samples to be investigated while the entire surface remains in focus. To understand the principle of electron microscopy, let us consider the imaging resolution. The wavelength of the moving electron is given by

$$\lambda \left(nm^{-1} \right) = \frac{h}{p},$$ (4.3)

where h is Planck's constant and p is the particle's relativistic momentum. For an electron beam with energy E (eV), the wavelength is

$$\lambda = \frac{1.225}{\sqrt{E + 10^{-6}E^2}}.$$ (4.4)

The energy range of typical electron microscopes is 10^3–10^6 eV and, therefore, λ is in the order of picometers, smaller than the atomic diameters. This fact means that electron beams with proper energy would allow imaging even single atoms. The resolution of the electron microscope is also dependent on the imperfection of electron optics that limits the high-resolution imaging. Electron microscopy imaging requires sample preparation, and imaging is performed under high-vacuum. The samples should be stable under vacuum and electron irradiation conditions.

Electron microscopy has not only the ability to image the surface of the samples but also allows investigating electron diffraction that provides crystallographic information of the sample. The electron beam generates internal excitations in the samples' atoms and emission of characteristic X-rays. This feature allows investigating the elemental composition of the specific region, using the EDS method described in Chapter 2. Electron energy loss spectroscopy (EELS) can also be applied to investigate the localized elemental composition of samples, when using sufficiently thin samples.

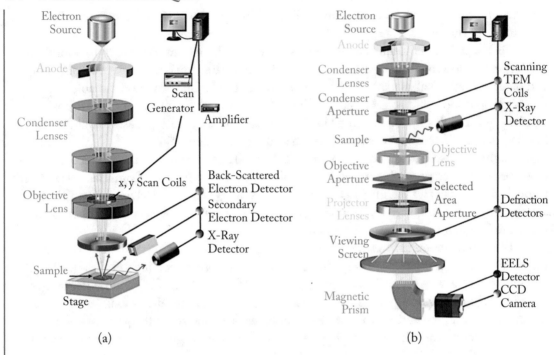

Figure 4.6: Schematics of SEM (a) and TEM (b) microscopes.

Depending on the configuration of the electron beam, the sample, and the detectors, electron microscopes can be classified into two groups. In a scanning electron microscope (SEM), the detectors are mounted on the same side of the incoming electron beam (Figure 4.6a). The electron probe microanalyzer (EPMA) is another microscopy method, designed to analyze characteristic X-rays to determine the elemental composition of samples. In transmission electron microscopy (TEM) most detectors are mounted behind the sample, and electrons are transmitted through the thin samples (Figure 4.6b). This feature allows high-resolution imaging of the internal details of the samples.

In SEM, the electron beams, having $\sim 0.01–1$ μm diameter, are focused on the surface of the sample, which is scanned by small areas (pixels). The signal of these pixels is recorded individually. Magnification is determined by the ratio of the size of the displayed image to the scanned area. The electron beam diameter and interaction volume with the sample determine the imaging resolution. In TEM, magnification is achieved using a set of lenses underneath the sample.

An essential aspect of electron microscopy imaging is obtaining a signal that contributes to correct contrasts. This is primarily related to the interaction of the incoming electron beam with the sample. Most of the incident electrons strongly backscatter and lose their energy. Interaction

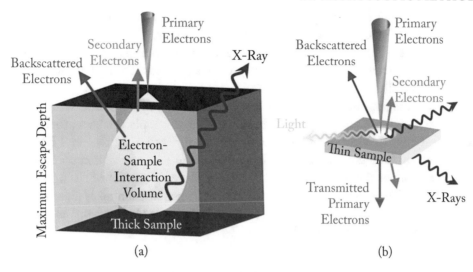

Figure 4.7: Interaction of electron beam with SEM (a) and TEM (b) samples, depicting effects of interaction volume.

of incident electrons with the sample also generate emission of secondary electrons, photons, and X-rays from the sample (Figure 4.7a). In thin (below 100 nm) TEM samples, the backscattering of primary electrons is minimized (Figure 4.7b). In SEM imaging, lateral resolution is strongly dependent on the volume of the sample interacting with the incident electrons. In EELS and electron diffraction methods, the scattering resolution is similar to the probe size because these techniques use elastic (or nearly elastic) scattered electrons.

In the case of SEM imaging, the samples can be sufficiently thick and incident electrons are stopped in the sample, due to multiple scattering events. Such interaction creates significant internal excitations and results in the generation of secondary electrons from the surface of the sample. Some portion of the primary electrons pass closer to the atomic nucleus of the sample material and are scattered out in a large angle. Another portion of the incident electrons may be subjected to multiple scattering events and emerge, back from the sample surface. Those backscattered electrons can also be used for imaging. The kinetic energy of scattered electrons may remain unchanged after the scattering events (elastic scattering). When the electron loses its energy, the scattering is said to be inelastic.

In SEM imaging, secondary electrons are used for imaging. These electrons have mostly exited from the valence levels of the sample's atoms, and some are primary electrons subjected to multiple inelastic scattering events. Secondary electrons are concentrated close to the primary electron beam spot and serve as the main SEM imaging method, providing topographic contrast. High-energy backscattered electrons provide an excellent opportunity for atomic number contrast imaging and crystallographic information. The lateral resolution ($\sim$ 100 nm) of backscattering electron imaging is relatively low compared to secondary electron imaging. This is related to the fact that high-energy backscattered electrons can escape from deeper layers of the sample surface (Figure 4.7a). The coefficient of backscattering electrons increases with the atomic number. Therefore, phases containing elements with sufficiently different atomic numbers can be distinguished easily.

In TEM imaging, most of the incident electrons pass through thin samples, since the mean distance between scattering events is similar to the sample thickness (below 100 nm). Elastically scattered electrons are used to measure the diffraction pattern in TEM. Electrons passed through the thin sample can provide a large amount of information on crystallography and elemental composition of artworks and artifacts. When the beam of electrons is scattered from a sample with a crystalline structure, interference phenomenon takes place that is similar to X-rays, as described in Section 2.6. The scattered intensity is observed at θ angles, according to the Bragg's law (see Equation (2.5)):

$$\theta = \arcsin\left(\frac{\lambda}{2d}\right), \tag{4.5}$$

where d is the interplanar distance between lattice planes of crystalline solids and λ is the wavelength. Since λ is small, θ also has small values (below 0.5°), therefore,

$$\theta = \frac{\lambda}{2d}. \tag{4.6}$$

The scattered electrons emit through a thin sample and then pass through the electromagnetic lens. The lens focuses the electrons leaving the sample at a particular angle to its focal point. If the sample has a single crystal that is larger than the beam diameter, and the lens is set to focus this plane onto the detector (fluorescent screen), a spot diffraction pattern is observed (Figure 4.8a). When the sample is polycrystalline on the scale of the beam, a ring (or superposed spot) pattern is formed on the detector (Figure 4.8b). It should be noted that only atomic planes that are parallel to the electron beam will have a significant contribution to the diffraction pattern. Spot diffraction patterns can be considered a scale map of the reciprocal crystalline lattice of the sample. The actual distance between the spots is inversely propositional to the spacing between the atomic planes producing them. Knowing the *camera constant*, the effective magnification of the system, one can identify the crystal structure and the interatomic spacing from the diffraction patterns. This method is rarely applied in the cultural heritage field, most likely due to the lesser availability of TEM microscopes and the difficulties associated with sample preparation.

The sample preparation in optical or electron microscopy is a critical step, and depends on the type of information being sought. In some cases, minimal or sometimes no sample preparation is needed. Small samples taken from the bulk of the cultural heritage object are often immersed in epoxy resin. After solidification, polishing in sequential steps is performed with abrasive papers having grain sizes from 180–4000 mesh. Then a polishing cloth is used with

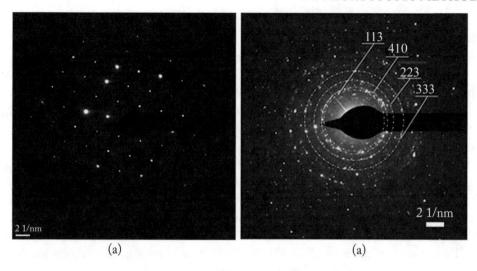

Figure 4.8: Spot (a) and superposed spot (b) electron diffraction patterns.

diamond or other abrasive paste. Metallic objects with smooth surfaces provide little or no contrast for OM imaging. The samples can, therefore, be etched with chemicals. The solutions of these chemicals preferentially attack some regions to leave an uneven surface. In this way, the microscope reveals the microstructure of crystalline materials, such as the different phases present in an alloy. The etchant solutions dissolve the regions between the individual crystallites (grains) of the samples, leaving a grain-boundary groove that is visible as a dark line. In SEM imaging, the electrically insulating sample (such as ceramics) should be coated with a thin layer (few nanometers) of conductive material (e.g., carbon) before imaging. This method allows for avoiding a periodic charging-discharging process that decreases the image quality. Electrically conductive samples immersed in insulating epoxy should also be coated with such a layer, to provide efficient dissipation of electrons through the SEM's sample holder.

TEM samples should be electron transparent (50–100 nm), and the preparation of such samples is an elaborate process. Depending on the material, the method for making the TEM sample is varied. The steps include the removal of unwanted material by chemical or mechanical methods and cutting of the sample to the desired thickness with different methods, such as ion beam milling. This sample preparation can be performed in a dual-beam (electron-ion) microscope. In some cases, small objects of interest, such as coins, can be inserted into a microscope and micrometer-sized samples can be extracted from it. Then samples can be polished inside the microscope by a focused ion beam down to 50–100 nm thickness.

4.4.2   EXAMPLES OF MICROSCOPIC IMAGING

Imaging Bronze Objects

Bronze is an alloy that contains primarily copper and tin. The use of copper-tin bronze in the ancient world dates back to the late fourth millennium. The Iranian Plateau, rich in copper ores, became one of the earliest bronze metallurgy centers in the world. Investigations on the microstructure of these artifacts allow for determining the manufacturing procedures of bronze from this region. This information also enables revealing the corrosion process that led to the degradation of the objects.

A metallographic sample of an Iranian bronze object [6] examined under the OM before etching shows the alloy matrix with numerous scattered dark inclusions that are mostly elongated (Figure 4.9). These inclusions, residue of initial ores, appear as a grey-green color in the image. In the bronze smelting process, copper sulfide or oxide ores mixed with tin ore were smelted in the crucible to produce bronze. Another method was also used, to add the tin ore (SnO_2) to molten copper. The presence of these sulfur-containing dark inclusions suggests extensive use of sulfide ores, and their elongated shape indicates intensive processing, such as hammering. Corrosion layers formed on the surface of the samples can also be seen. To reveal the grains' microstructure, samples can be etched using a solution. After this process, the images show recrystallized grains of the copper-tin solid solution, with twinned and slip lines within the grains. Figure 4.10 shows the optical microscopy image of a small etched sample taken from a bronze bowl discovered in Iran, tentatively dated back to the first century BC [7]. This image shows thermal twin bands and slip lines between alloy grains. The twinned grain evidences that the cast bronze ingot was first produced and then shaped into a thin metal vessel by multiple mechanical working (hammering) and annealing steps. The hammering causes the metal sheet to work-harden.

A heat treatment at 500–800°C was likely used between hammering steps to extend the workability of copper alloys. The non-homogeneous grain size and residual slip lines between the grains suggest that the annealing step was performed for a short time or at low temperatures. Such processing did not provide a complete recovery of the structural defects. A longer annealing time or higher temperature would facilitate the recrystallization process that would have allowed the formation of grain with approximately the same sizes. Figure 4.10a also shows shrinkage cavities and inclusions in the sample.

The cross-sectional OM image of a sample taken from a corroded part of the bowl shows that the thickness of the patina layer is $\sim 150~\mu$m (Figure 4.10b). The dark inner layer (marked with #1) is the alloy, while the reddish layer (#2) identified (using Raman spectroscopy) as a mixture of cuprite (Cu_2O) with cassiterite (SnO_2). Next to this cuprite/cassiterite layer is the pure cuprite (#3), followed by a malachite layer (#4). The corrosion of bronze objects in soil depends on the nature of the alloy and the method of object manufacture, soil composition, humidity, porosity, and other factors, which lead to the formation of layered patina. This cross-sectional imaging suggests that the SnO_2-rich layer is defective and discontinuous. Therefore, slow dif-

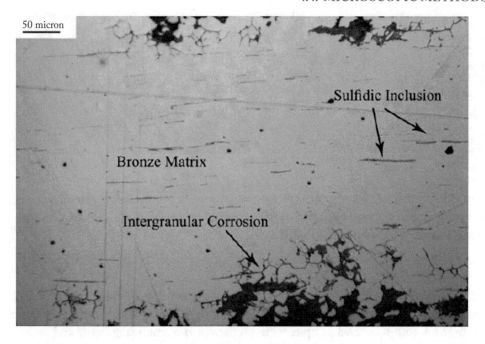

Figure 4.9: Optical microscopy image of an ancient Iranian bronze, before etching. Reprinted with permission from [6]. Copyright 2014 Elsevier.

fusion of copper through this layer to the surface converted it into the malachite (identified by the Raman spectroscopy) layer.

Imaging Gilded and Silvered Objects

Mercury-gilding and silvering are well-known age-old approaches, whereby a thin layer of precious metal is applied on the surface of metallic artifacts (such as copper). In these methods, jewels, statues, artworks, ornaments, and other objects were first coated with amalgam paste. These pastes were prepared by mixing gold or silver with mercury. A thin layer of paste was applied on the clean metal substrate. Thermal treatment of coated samples at 200–300°C allowed a large quantity of the mercury to slowly evaporate. Further thermal processing at higher temperatures removed mercury residues. This process produced a porous layer of silver or gold with a thickness from 1–10 μm. The final burnishing was performed mechanically, using a bone or agate tool to reduce the porosity and produce objects with shiny metallic color. The long exposure of these objects to different degradation processes during soil burial, in outdoor or indoor atmospheric interactions, changed their surface morphology and chemical composition. During such long interaction, corrosion products were formed with specific characteristics that are different from the degradation products grown on archaeological or historical artifacts. For ex-

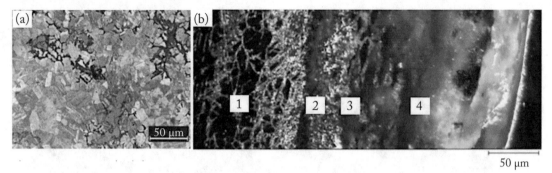

Figure 4.10: Optical images of grain structure and non-metallic inclusions on sample surface (a) and cross section of the corrosion patina (b), observed on the side of the bowl. Reprinted with permission from [7]. Copyright 2014 Elsevier.

ample, microscopic observations of the silvered and gilded objects found during the cleaning operations at the banks of the Tiber River (Rome, Italy) uncovered some interesting corrosion processes [8]. Cross-sectioned samples were prepared from small fragments and embedded in epoxy, then micro-sectioned to preserve the chemical and structural features of the outer layers. After careful polishing, mirror-like surfaces were examined using both electron and optical microscopes.

Such imaging revealed a complex morphology and surface chemical composition of the coated artifacts. EDS microanalysis showed the presence of a conspicuous (7–9 wt%) amount of mercury in the gilded layer, confirming the use of the amalgam gilding method to produce the coating. The cross-sectional OM (Figure 4.11a) and SEM (Figure 4.11b) images with backscattering electrons show that the thickness of the gold layer is less than 10 μm. EDS microanalysis does not show any Au and Hg in the substrate under the gold layer, which has a granular structure and porous areas. Figure 4.11c shows that this layer mainly contains copper and oxygen, islands of Pb-Sb, as well as other light elements such as chlorine and phosphorous. The presence of O, P, and Cl can be explained by the interaction of the copper substrate with water-soluble species from the soil, during the burial.

Backscattered electron SEM imaging (Figure 4.11b) revealed that the gilded layer has multiple cracks due to micro sized manufacturing defects. These defects in the gold layer played a damaging role during burial, when the water-soluble species in the soil came into contact with the copper substrate. This process created galvanic areas and resulted in the formation of a corrosion layer consisting of primarily Cu_2O. The expansion of the corrosion layer caused the breaking and detachment of some areas in the gilded layer (Figure 4.11d).

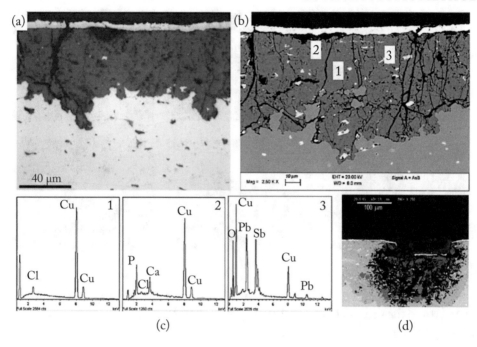

Figure 4.11: OM (a) and backscattering SEM (b) images; EDS spectra (c) of corroded copper substrate, transformed in cuprite (Cu_2O), with local Pb-Sb islands; and SEM image (d) of corrosion induced rupture of gilded layer. Reprinted with permission from [8]. Copyright 2016 Elsevier.

Imaging Silver Coins

Electron microscopy methods have been used to investigate the microstructure of ancient Ag-Cu alloys. Ag-Cu alloys belong to so-called eutectic type systems, and the solubility of Ag in Cu and Cu in Ag is limited. The solubility of one constant into another decreases as the temperature drops. There is one composition (28%Ag–72%Cu) at which the liquid melt directly solidifies. At other compositions, liquid-solid equilibrium exists in the broader temperature range. Solidification of melt in these alloys produces two distinctive phases: Ag-rich (α-phase), and Cu-rich (β-phase).

Metallographic investigation of polished Ag-Cu alloys by SEM allows for imaging these two phases [9]. For example, polished silver-copper alloy imaged by backscattering electron and EDS elemental mapping methods reveals the Ag-rich-layer phase, with a lighter contrast (higher energy backscattered electrons) to silver, and the grey phase is distinguished from the Cu (Figure 4.12a). The backscattering electron image shows that the grains of the β-phase are surrounded by the mixture of both α and β phases, containing, on average, 72% silver. This image also suggests that dendritic growth is the primary segregation mechanism that occurred

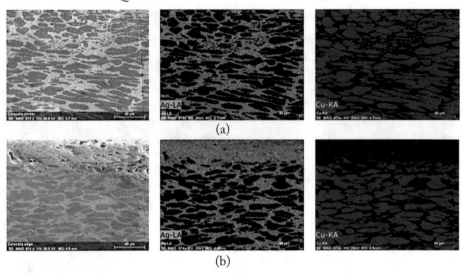

Figure 4.12: SEM images with backscattered electrons and EDS elemental distribution images of the core (a) and the surface (b) of a polished Ag-Cu alloy. Reprinted with permission from [9]. Copyright 2019 Elsevier.

during the casting of these alloys. Other processes that can happen during the solidification of Ag-Cu melts are normal and inverse segregation. In the normal segregation process, the lower melting point phase is concentrated toward the mold core. In inverse segregation, the Ag-rich phase can concentrate on the outer surface of the molten alloy just before solidification. Inverse segregation takes place in slow cooling conditions, which probably were not the case in ancient silver mints. Metallographic investigation of the surface of ancient Ag-Cu alloys often reveals an enriched surface layer with pure Ag (Figure 4.12b). The thickness of the Ag-rich layer in this particular case is ~ 30 μm. EDS microanalysis of this surface layer shows $\sim 95\%$ silver.

This complex structure has been achieved by preparing a Cu-rich alloy, followed by short heating in air to oxidize Cu at the surface. Then the oxidized alloy was soaked in vinegar to dissolve-out Cu from the surface. This process formed a copper-depleted (Ag-rich) porous layer on the surface of the alloy. The burnishing of this layer restored the silver-metallic luster of the alloys while using significantly less silver in the production. This approach was prevalent in the preparation of silver coins in the ancient world.

The destructive metallographic imaging shown in Figure 4.12 is not acceptable for artifacts with high historical value. Focus ion beam (FIB) assisted SEM imaging coupled with EDS microanalysis allows examination of the microstructure and composition of such samples [9]. This is a micro-distractive method and uses an energetic ion (Ga) beam to mill microscopic trenches on the metal surface. This method is a proper approach for analyzing small objects.

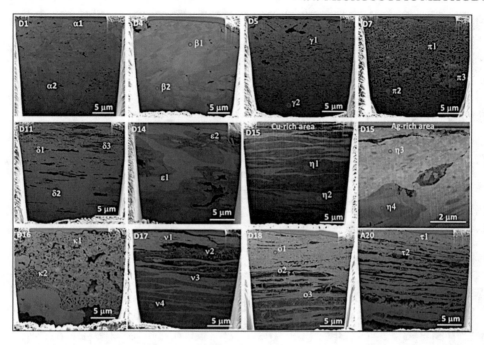

Figure 4.13: SEM images of surface morphology of Roman coins prepared by FIB-assisted SEM imaging. The elemental composition of analyzed areas by EDS are marked with Greek letters that can be found in Table 4.2. Reprinted with permission from [9]. Copyright 2019 Elsevier.

A set of SEM-EDS images for the Roman Republic and Imperial coins investigated by this method are presented in Figure 4.13 and Table 4.2.

Figure 4.13 shows cross-sectional images of several Roman Republic coins. D1 (136 BC) and D4 (30 AD) exhibit large (2–5 μm) grains and small pores (0.1–0.5 μm). EDS elemental microanalysis results (Table 4.2) in the $\alpha 1$, $\alpha 2 \beta 1$, and $\beta 2$ areas suggest that Ag content is $\sim$ 98%, with some Au, Pb, Sn, Mg, Al, Ti, Si, O, and Cl impurities. The D5 (65 AD) and D7 (107 AD) coins have irregular and slightly elongated pores; the leaching of Cu created these elongated pores, suggesting that the coin alloys were probably subjected to hammering. EDS elemental analysis results for coin D5 indicate 95% Ag, 1.5% Cu, 1% Au, and some lighter elements. Coin D7 contains a significant amount of sulfur at areas closer to the surface, with 96% Ag as a primary element.

The SEM image of coin D11 (178 AD) shows significantly directional pores and several deformed particles. EDS analysis (area $\delta 3$) suggests that these particles are composed of silicon and oxygen. Such composition indicates that the particles could be byproducts attached to the surface of the coin alloy, which deformed and broke into smaller pieces and embedded into the alloy during mechanical working. The alloy consists of 90–95% Ag, Cu, some Au, and Pb, as well

Table 4.2: EDS analysis results for the areas marked with Greek letters in Figure 4.13. Experimental error is 1.0–2.0%.

Coin Number	Area Market on Figure 7	Ag	Cu	Au	Pb	Sn	Mg	Al	Ti	Fe	Hg	Si	O	S	Cl	P
D1	α1	97.68		1.23			0.26	0.26				0.22			0.35	
	α2	98.08		0.13	0.13	0.44	0.39	0.28	0.14			0.13			0.28	
D4	β1	97.36		1.15			0.67	0.22	0.4			0.08	0.12			
	β2	97.96		0.96		0.37	0.22					0.09	0.15		0.25	
D5	γ1	95.76	1.53	1	0.11	0.38	0.46					0.11	0.31		0.34	
	γ2	95.62	1.64	0.8			0.24		0.43			0.03	0.38	0.57	0.29	
D7	π1	89.36	3.73				0.46	0.43					0.87	4.59		0.56
	π2	95.67	1.56	0.94			0.38		0.41				0.59	0.09	0.27	0.09
	π3	83.56	3.88	9.28		0.48	0.9		0.21				0.59	0.9	0.11	0.09
D11	δ1	94.4	2.81		0.83		0.56	0.16	0.32	0.61		0.13	0.18			
	δ2	90.94	4.68	1.03	0.22		1.58	0.2	1.01			0.06	0.28			
	δ3	71.32	2.95		0.25	0.18	0.53	0.15	0.11	0.64		17.1	6.77			
D14	ε1	85.31	11.8	1.19	0.4		0.33	0.24	0.47			0.03	0.23			
	ε2	20.1	71.29	2.45			0.58	0.07	0.15	0.79		0.04	4.53			
D15	η1	10.71	86.99	1.92					0.23				0.15			
Cu-rich area	η2	82.06	15	1.18	0.51	0.27	0.39	0.11				0.34	0.14			
D15	η3	82.03	3.72	0.31	0.21						13.73					
Ag-rich area	η4	13.41	80.97	1.59			0.66	0.05	0.26			0.04	3.02			
D16	κ1	98.06	0.48	0.88			0.05	0.14	0.34			0.4	0.13			
	κ2	98.07	0.22	0.76		0.78	0.34	0.14	0.34			0.04	0.09			
	ν1	90.95	6.1	0.79			0.62	0.29	0.19			0.67	0.39			
D17	ν2	47.19	43.17	1.94	0.17	0.37	0.47	0.02					6.36	0.31		
	ν3	87.99	8.94	0.91	0.18		1.05	0.18	0.24					0.51		
	ν4	16.14	77.7	1.75			1.24		0.14				3.03			
D18	o1	95.29	2.48	0.3			0.31	0.16				0.38	0.3			
	o2	69.09	2.89	1.08			0.88	0.13	0.31			19.62	6			
	o3	60.41	29.75	3.41			2.81	1.01				0.89	1.72			
A20	τ1	97.07	1.15	0.31			0.5	0.11	0.37	0.24		0.06	0.19			
	τ2	85.23	3.52	0.64	0.17		0.45		0.24			7.04	2.71			

as several lighter elements. D14 coin (199 AD) shows two distinct Ag-rich (light) and Cu-rich (darker) areas.

Imaging and EDS analysis of coin D15 (210 AD) were performed in two different areas. The picture taken closer to the center of the coin shows Cu- and Ag-rich phases. The grains of these phases are significantly deformed due to intense hammering. In the coin edge, a surface layer exhibits fine, brighter non-deformed grains with darker enclosures. EDS analysis (point $\eta3$, Table 4.2) revealed light grains with $\sim$ 80% silver and $\sim$ 15% mercury composition, with some minor elements. The main component of the phase with a darker contrast is Cu. These results suggest that the mercury-silvering was probably used for developing an Ag surface layer on the copper-rich Ag-Cu alloys.

The coin D16 (213 AD) has a highly porous surface. EDS analysis ($\kappa1, \kappa2$ areas, Table 4.2) discloses 98% silver. The porous nature and high Ag content suggest that there was selective leaching of copper. The SEM image of D17 (220 AD) coin also shows chemical enrichment of the surface by copper leaching. The surface of the Ag-rich layer exhibits low porosity. Few pores and silver enriched areas ($\nu2$) are seen beneath the Ag-rich layer. These pores led to the local fracture of the surface layer during the mechanical processing of the coin. Below the surface layer, the coin has two-phase morphology with Ag- and Cu-rich elongated grains. D18 (224 AD) and A20 (240 AD) coins' surfaces are also significantly depleted with Cu. The silver content in this layer (areas $o1$ and $\tau1$) is above 95%, suggesting that the coins were subjected to Ag surface enrichment.

TEM Imaging of Artworks and Artifacts

An excellent example of using TEM methods in cultural heritage analysis is the investigation of the glazes of ceramic pottery [10]. Some medieval glazed pottery contains a metallic lustre, which is a thin film formed just below the surface. These films were made by mixing some metal (Ag, Cu) salts with vinegar and lye (sodium hydroxide obtained by leaching ashes). This mixture was deposited on the glaze and subjected to heat treatment. The burning of organic residues in the furnace reduced the metal ions to metallic nanoparticles and dispersed them on the near-surface glaze layer.

Figures 4.14 TEM analysis for two samples of such medieval glazed pottery. The first sample was excavated from Fustat (the first capital of Egypt under Muslim rule). The second sample is a sherd (broken piece of ceramic material) excavated from Termez (a city in Uzbekistan destroyed by Tamerlan in the 4th century). Small, polished TEM specimens were taken from both objects. A set of TEM images, for both materials, display a similar general microstructure (Figures 4.14a and b). Both samples exhibit a darker layer of near-spherical particles embedded in a transparent amorphous matrix. The nanoparticle layers for both samples are about 20–80-nm thick. Each layer exhibits a density gradient.

EELS and electron diffraction analyses were used to determine the composition of metal particles. Most of the particles in the Fustat samples were pure silver, while particles in the

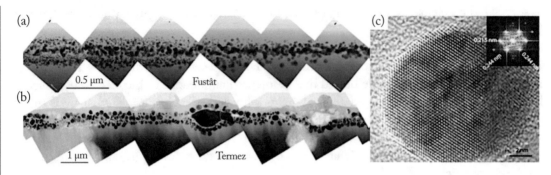

Figure 4.14: Two sets of TEM images for glazed pottery samples excavated from Fustat (a) and Termez (a) as well as atomic level high-resolution TEM images of Ag nanoparticles in the Fustat sample, as observed in (110) zone axis particle (c). Reprinted with permission from [10]. Copyright 2008 Taylor & Francis.

Termez sample were almost pure copper. Electron diffraction patterns show the presence of some cuprite in the Termez sample. High-resolution TEM images of a silver nanoparticles in the Fustat sample (Figure 4.14c) show quasi-spherical shape along the (110) zone axis (determined by Fast Fourier Transform of the TEM image as an inset in Figure 4.14c). Electron microscopy investigations allow for determining the morphology and structure of pottery samples and help in reconstruction efforts. These investigations suggest that nanotechnology existed in medieval times. The masters were able to incorporate high-quality metallic nanoparticles beneath the surface of amorphous materials.

Modern TEM methods also allow an unprecedented examination of the composition of paint layers. In standard SEM-EDS investigation, the resolution of EDS elemental analysis is 0.5–1 μm, due to the significant interaction volume of electrons with the sample. Thin lamella removed from specific areas of the samples, by the FIB method, enables high-resolution microscopic examination. This includes the utilization of scanning TEM (STEM) imaging of each lamella and elemental mapping with the EDS. In this TEM-based method, the resolution is closer to the size of the electron beam (below one nanometer). An example of such investigation is the revealing of the paint-layer composition of the iconic *Girl with a Pearl Earring* (Figure 4.15a) created by Johannes Vermeer in 1665 [11]. Vermeer applied a painting technique using sophisticated underlayers, followed by the application of top paint layers. To reveal the composition of these layers, cross-sectional lamella (Figure 4.15b) were taken from the *Girl's* jacket using FIB (Figures 4.15c–e) and investigated by STEM/EDS analysis (Figures 4.15f and g).

The results of these analysis suggest that a significant component of the ground layer was identified as chalk, based on the presence of large (1–2 μm) particles containing calcium (Figure 4.15f, in blue). A smaller amount of lead white can also be seen at the boundaries of

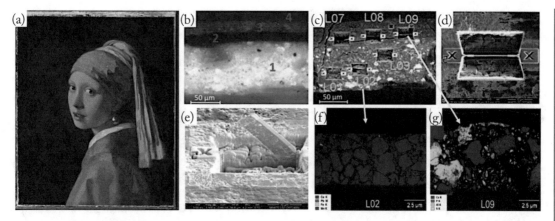

Figure 4.15: Microscopic imaging and EDS analysis of sample taken from the dark part of the jacket of *Girl with a Pearl Earring* by Vermeer (a). OM image of cross-sectional sample (b) numbered with ground layer (1), underlayer (2), paint layer (3), and varnish (4). Preparation of lamella at different positions of the sample by FIB and lifting out of a lamella (c, d, and e). STEM/EDS analysis mapping overlay of L02 lamella (f) taken from the ground layer: Ca (blue), Mn (magenta), Fe (green), Pb (red), and lamella L09 from the dark underlayer: P (red), Ca (blue), S (cyan), Al (green) https://heritagesciencejournal.springeropen.com/articles/10.1186/s40494-019-0308-4.

the chalk particles, identified by lead (in red). This technique, of the filling of voids in the calk layer with lead white, allowed preparing the ground layer with practically no pores. This method, of laying out the ground layer using pigments with different particles, is known as the Dutch stack process. The dark-underlayer elemental composition is the following: calcium, iron, sulfur, phosphorus, aluminum, as well as small amounts of lead, sodium, and potassium (Figure 4.15g). The black pigment is bone black mixed with a small amount of charcoal, as identified by the backscattered electron image. The fine aluminum phase is possibly an aluminum oxide-based lake pigment. Some particles in this layer contain sulfur and calcium in proportions that suggest gypsum.

4.5   REFERENCES

[1] L. Fragasso and N. Masini, Postprocessing of infrared reflectography to support the study of a painting: The case of Vivarini's, *Polyptych. Int. J. Geophys.*, 4(25):1–9, 2011. DOI: 10.1155/2011/738279. 164

[2] M. Gargano, A. Galli, L. Bonizzoni, R. Alberti, N. Aresi, M. Caccia, I. Castiglioni, M. Interlenghi, C. Salvatore, N. Ludwig, and M. Martini, The Giotto's workshop in the XXI

century: Looking inside the "God the Father with Angels," *Gable. J. Cult. Herit.*, 36:255–263, 2019. DOI: 10.1016/j.culher.2018.09.016. 166

[3] S. Zesch, S. Panzer, W. Rosendahl, J. W. Nance, S. O. Schönberg, and T. Henzler, From first to latest imaging technology: Revisiting the first mummy investigated with X-Ray in 1896 by using dual-source computed tomography, *Eur. J. Radiol. Open*, 3:172–181, 2016. DOI: 10.1016/j.ejro.2016.07.002. 169, 170

[4] S. Kichanov, I. Saprykina, D. Kozlenko, K. Nazarov, E. Lukin, A. Rutkauskas, and B. Savenko, Studies of ancient Russian cultural objects using the neutron tomography method, *J. Imag.*, 4:1–9, 2018. DOI: 10.3390/jimaging4020025. 171

[5] D. Mannes, E. Lehmann, A. Masalles, K. Schmidt-Ott, A. V. Przychowski, K. Schaeppi, F. Schmid, S. Peetermans, and K. Hunger, The study of cultural heritage relevant objects by means of neutron imaging techniques, *Insight*, 56:137–141, 2014. DOI: 10.1784/insi.2014.56.3.137. 172

[6] O. Oudbashi and P. Davami, Metallography and microstructure interpretation of some archaeological tin bronze vessels from Iran, *Mater. Charact.*, 97:74–82, 2014. DOI: 10.1016/j.matchar.2014.09.007. 178, 179

[7] C. Soffritti, E. Fabbri, M. Merlin, G. L. Garagnani, and C. Monticelli, On the degradation factors of an archaeological bronze bowl belonging to a private collection, *Appl. Surf. Sci.*, 313:762–770, 2014. DOI: 10.1016/j.apsusc.2014.06.067. 178, 180

[8] G. M. Ingo, C. Riccucci, M. Lavorgna, M. Salzano de Luna, M. Pascucci, and G. Di Carlo, Surface investigation of naturally corroded gilded copper-based objects, *Appl. Surf. Sci.*, 387:244–251, 2016. DOI: 10.1016/j.apsusc.2016.06.082. 180, 181

[9] K. Manukyan, C. Fasano, A. Majumdar, G. F. Peaslee, M. Raddell, E. Stech, and M. Wiescher, Surface manipulation techniques of Roman Denarii, *Appl. Surf. Sci.*, 493:818–828, 2019. DOI: 10.1016/j.apsusc.2019.06.296. 181, 182, 183

[10] C. Mirguet, P. Fredrickx, P. Sciau, and P. Colomban, Origin of the self-organisation of $Cu°/Ag°$ nanoparticles in ancient lustre pottery. *A TEM Study. Phase Transit.*, 81:253–266, 2008. DOI: 10.1080/01411590701514433. 185, 186

[11] A. Vandivere, A. van Loon, K. A. Dooley, R. Haswell, R. G. Erdmann, E. Leonhardt, and J. K. Delaney, Revealing the painterly technique beneath the surface of Vermeer's girl with a pearl earring using macro- and microscale-imaging, *Herit. Sci.*, 7(64):1–16, 2019. DOI: 10.1186/s40494-019-0308-4. 186

CHAPTER 5

Archaeological Dating

5.1 INTRODUCTION

Everything which has come down to us from heathendom is wrapped in a thick fog; it belongs to a space of time we cannot measure. We know that it is older than Christendom, but whether by a couple of years or a couple of centuries, or even by more than a millennium, we can do no more than guess.

(Rasmus Nyerup, 1802)

This is an amazing statement, barely imaginable today with the wealth of technology available for measuring time and probing the past. Dating the past is a human desire that came with the period of Enlightenment when the drive for scientific understanding overshadowed religious belief in biblical truth. Archbishop Usher (1591–1656) still relied on biblical time, as given in the chronicles of the Old Testament, to publish in 1634 the first day of creation as Sunday, October 23, 4004 BC. His declaration was only eight years before that of the Anglican clergy and Cambridge scholar John Lightfoot. He derived the day of creation to be September 17, 3928 BC, nearly 100 years earlier, as indicated on his calculation table in Figure 5.1. The difference in results suggests uncertainties in the input data and methodology of these early attempts to determine the ages of history, a general problem that still haunts us today using modern scientific methods and techniques.

While in the present, time can be measured with accuracy in pico-seconds and below, the measurement of age poses larger challenges since the quality of age determination depends on the uncertainties associated with the method. Every scientific method and tool, as well as the underlying theoretical assumptions, introduces its own uncertainties, which then translate into the uncertainty of the age determination. In many cases, considerable corrections or normalizations of the data are required to obtain a final age. These corrections often require tedious research and analysis of the history of the specific samples and frequently, also, of the specific reference frame. While the techniques are getting ever better and more sophisticated, the underlying assumptions remain the same and will, therefore, be a large part of the following discussions of archaeological dating methods.

5.2 DENDROCHRONOLOGY

Li circuli delli rami degli alberi segati mostrano il numero delli suoi anni, e quali furono più umidi o più secchi la maggiore o minore loro grossezza. (The rings around the branches of trees that have been sawn show the number of its years and their thickness depends on whether the year had been wetter or drier.)

(Leonardo da Vinci, 1452–1519.)

Since the days of Archbishop Usher, several methods have been developed to determine the age of archaeological and anthropological samples. Initially, the dating was based on an analysis of style, which led to a systematic categorization of styles based on the ornamental and artistic structure of the artifacts. Later, during the second half of the 20th century, a number of unique scientific methods were developed using analytical tools from biology, chemistry, and physics.

A classic, simple method for dating objects that were made from wood is dendrochronology (or tree-ring method) by which tree rings are simply counted to determine the number of years. For several kinds of trees, such as oak and firs, the width of the annual tree ring depends on the climate conditions during the year, the amount and distribution of rainfall and sunny days. In particular, a warm and humid spring causes the tree ring to grow faster. Through the variation of climate conditions during the lifetime of a tree, a characteristic sequence of tree rings will emerge, that represents the local climate over time. This information can be used to determine the age of a specific tree and to develop a scale for calibrating the age of wooden materials back into the distant past.

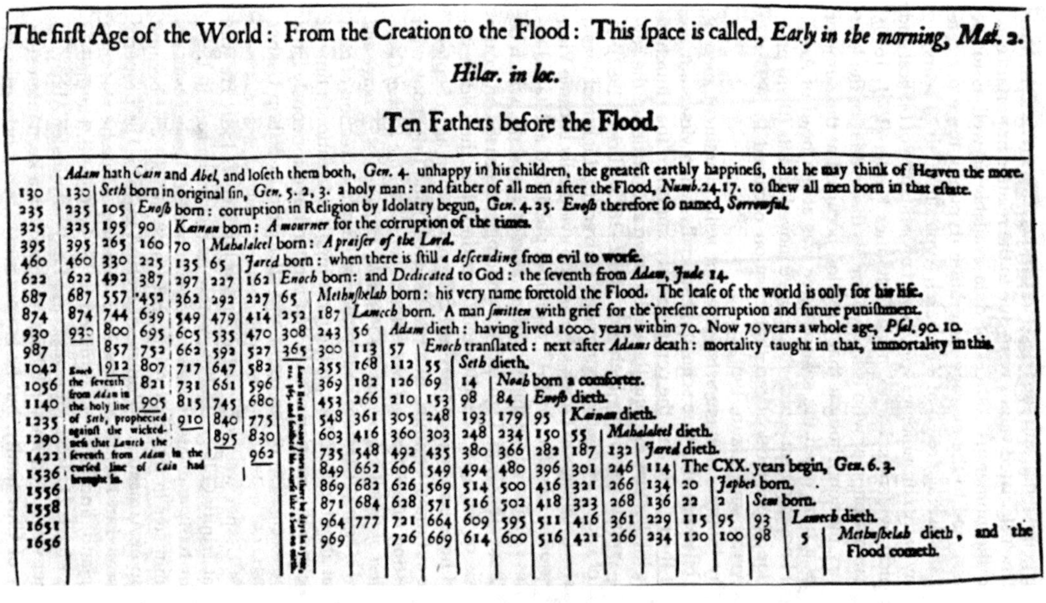

> **The first Age of the World: From the Creation to the Flood: This space is called, *Early in the morning, Mat. 2.***
>
> *Hilar. in loc.*
>
> **Ten Fathers before the Flood.**
>
> *Adam* hath *Cain* and *Abel*, and loseth them both, *Gen.* 4. unhappy in his children, the greatest earthly happiness, that he may think of Heaven the more.
>
> *Seth* born in original sin, *Gen.* 5. 2, 3. a holy man: and father of all men after the Flood, *Numb.* 24. 17. to shew all men born in that estate.
>
> *Enos* born: corruption in Religion by Idolatry begun, *Gen.* 4. 25. *Enos* therefore so named, *Sorrowful.*
>
> *Kainan* born: *A mourner* for the corruption of the times.
>
> *Mahalaleel* born: *A praiser of the Lord.*
>
> *Jared* born: when there is still *a descending* from evil to worse.
>
> *Enoch* born: and *Dedicated* to God: the seventh from *Adam, Jude* 14.
>
> *Methushelah* born: his very name foretold the Flood. The lease of the world is only for his life.
>
> *Lamech* born. A man *smitten* with grief for the present corruption and future punishment.
>
> *Adam* dieth: having lived 1000. years within 70. Now 70 years a whole age, *Psal.* 90. 10.
>
> *Enoch* translated: next after *Adams* death: mortality taught in that, immortality in this.
>
> *Seth* dieth.
>
> *Noah* born a comforter.
>
> *Enos* dieth.
>
> *Kainan* dieth.
>
> *Mahalaleel* dieth.
>
> *Jared* dieth.
>
> The CXX. years begin, *Gen.* 6. 3.
>
> *Japhet* born.
>
> *Sem* born.
>
> *Lamech* dieth.
>
> *Methushelah* dieth, and the Flood cometh.

Figure 5.1: John Lightfoot's Old Testament Chronology, published in 1642, determining the day of creation to be September 17, 3928 BC based on the chronology of the Old Testament.

For determining age and building a chronology, the specific characteristics of the tree ring sequences are aligned with the sequence found in calibration samples of well-known ages, which have been obtained by other dating methods. These samples may include wooden bars or planks from ships or buildings of a specific historic period. These samples allow for matching across different ranges of history, from medieval times back to the early Egyptian and Sumerian cultures, as demonstrated in Figure 5.2. There are also local differences to consider since trees are species that are subject to local climate conditions, which are reflected in the ring structure.

The actual method is straightforward; special drills are used to take long samples from the tree or the wooden plank or bar. The number of rings can be counted along the samples. The art is more in establishing a scale by comparing the sequence of tree rings from the present with sequences from the past. Dendrochronology of single trees does not, however, necessarily reflect global climate conditions and, therefore, care must be taken to analyze and interpret the sample within the framework of the local climate environment and extrapolate beyond that, by comparison with other samples, to obtain a normalization network for extrapolating toward a global age scale.

The importance of dendrochronology, in the context of this book, is not so much in its usefulness as a tool for age determination but in its ability to provide a scale, to which all other methods can be normalized. This is particularly important for ^{14}C dating since the ^{14}C abundance in our atmosphere fluctuates, depending on the environmental conditions associated with the earth's magnetic field, the activity level of our sun, and, in more recent times, on anthropological activities such as the burning of fossil fuel and the nuclear bomb test program, as outlined in the following chapter.

Determining the ^{14}C abundance in tree rings over an extended period of time gives a calibration curve for the annual variations of atmospheric ^{14}C over the same time period. The oldest trees on earth are the California bristlecone pines, which can reach an age of up to 8,000 years. They grow in a rather arid climate, at high altitudes of 3,000 m, in the White Mountains of California where they are shielded by the Sierra Mountain Range from western pacific wind and rainstorms. The conditions warrant their longevity, and the wood does not rot away in the local arid conditions, so tree ring analysis can be performed, which covers several thousand years. Since the ^{14}C budget in the atmosphere is global, this method provides a reliable calibration for the fluctuations and variations of atmospheric radiocarbon, as shown in Figure 5.3.

5.3   RADIOCARBON DATING

5.3.1   GENERAL CONSIDERATIONS

Radiocarbon dating was introduced as a new tool for age determination in the early 1950s by Willard F. Libby. It is based on the radioactive decay of ^{14}C, which has a known half-life of 5730 ± 40 years. This half-life is a constant number, not affected by environmental or other impacts. Its application is limited to biological material that is subject to carbondioxid CO_2 exchange with the atmosphere.

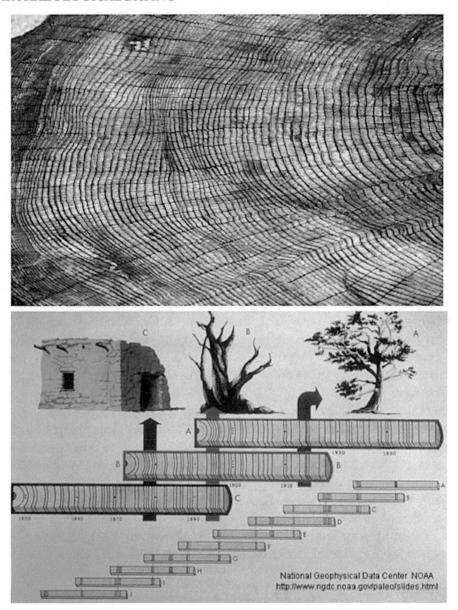

Figure 5.2: The growth rings of an unknown tree species, at Bristol Zoo, Bristol, England. The age of the tree at the time it was felled is determined by the number of rings in the trunk (https://www.ncdc.noaa.gov/data-access/paleoclimatology-data/datasets/tree-ring). A chronology over several tree generations can be established by comparing the ring sequences of tree samples from different times, as demonstrated in the figure. The ring sequences reflect the climatic conditions the tree has lived through and are therefore a primarily local measure and not a global one.

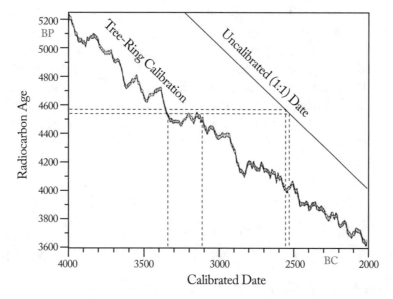

Figure 5.3: The calibration curve for the radiocarbon method, comparing the known age of a sample with the age determined by the amount of radiocarbon in the tree ring. The deviations reflect the fluctuations of ^{14}C over the years. Reprinted with permission from [7]. Copyright 2000 Elsevier.

The radiocarbon content in the atmosphere is due to the permanent bombardment of earth by an intense flux of cosmic rays, as outlined in the first chapter. Free neutrons, produced by the cosmic ray spallation processes, interact with the 14N content in the atmosphere through the ^{14}N(n, p)^{14}C charge exchange reaction, producing free ^{14}C atoms that nearly instantaneously combine chemically with atmospheric oxygen forming CO and CO_2 molecules, which become part of the biological carbon cycle. Through the carbon cycle or photosynthesis, the CO_2 and also $^{14}CO_2$ are imbedded in all plant materials. Atmospheric ^{14}C abundance depends on the intensity of cosmic ray bombardment and the ^{14}N content of the atmosphere but, on average, one can estimate that about 5 kg of ^{14}C is annually added to our atmosphere. This is an appreciable amount of radioactivity, which corresponds to nearly 83 MBq annually. Variation in the cosmic ray flux—due to changes in solar activity or also, in the strength and direction of the earth's magnetic field—will translate directly into a change in the radiocarbon content of the atmosphere. This can be directly observed in the ^{14}C fluctuation of the tree-ring carbon content.

Due to the production of ^{14}C by cosmic rays in the atmosphere, there is always a certain amount of radiocarbon embedded in all biological samples from plants to humans. These abundances establish themselves by way of the various exchange mechanisms that living organisms have with their outer environments, through the carbon cycle of plants, through breathing and digesting by animals and humans, and through absorption in ocean and other water surfaces. This results in an equilibrium of ^{14}C absorption and ^{14}C deposition being established in the organism. The moment the organism dies, the exchange ceases to operate; no carbon is absorbed and no carbon is deposited, except for decomposition processes that release CO_2 into the atmosphere. This decay process does, however, change the ^{12}C/^{14}C abundance ratio in the body material. The ^{12}C fraction in the decaying organism remains constant while the radioactive component ^{14}C decays gradually away. The change in the ^{12}C/^{14}C ratio corresponds to the time that has passed since death as derived by the decay equation shown below:

$$N(t) = N_0 \cdot e^{-\lambda \cdot t} \quad \Rightarrow \quad \ln \frac{N(t)}{N_0} = -\lambda \cdot t \quad \Rightarrow \quad \ln \left(\frac{N(t)}{N_0} \right)^{-1} = \lambda \cdot t$$

$$t = \frac{1}{\lambda} \cdot \ln \left(\frac{N_0}{N(t)} \right) = \frac{1}{\lambda} \cdot \ln \left(\frac{N(^{14}C)_{t=0}}{N(^{14}C)_{t=t}} \right) = \frac{1}{\lambda} \cdot \ln \left(N(^{12}C)_{t=t} \cdot \frac{R}{N(^{14}C)_{t=t}} \right) \quad (5.1)$$

$$\text{with } N_0 = N(^{14}C)_{t=0} = R \cdot N(^{12}C) \qquad R = \frac{N(^{14}C)_{t=0}}{N(^{12}C)_{t=0}}.$$

Here t is the age of the sample (starting from $t = 0$, the time the organic object died and stopped absorbing ^{14}C from the atmosphere), N(^{14}C) are the number of ^{14}C atoms, which corresponds directly to the β-decay activity of the sample as outlined earlier. N(^{12}C) is the number of ^{12}C atoms.

One has to determine the ratio of the ^{12}C/^{14}C content in the sample and, for calculating the age, one has to know the initial ^{12}C/^{14}C ratio in the atmosphere at the time the organism perished. If that ratio has declined to 50% of its initial value, the age of the sample corresponds to the half-life of ^{14}C. A classic example is the 1952 radiocarbon dating of the mummy of the Egyptian pharaoh Ramesses II (1303–1213 BC) by Willard Libby; in 1952 the mummy would have been 3,163 years old and the initial ^{14}C content of the body would have been decayed to 69% of its initial value (see Figure 5.4). However, extracting the age from the ^{14}C content is a bit more challenging: first, the amount of ^{14}C needs to be measured with high precision and second, the variations in the ^{14}C content in the atmosphere need to be understood and established with high accuracy over the here-considered age range.

In the following sections we will first present a short overview of the various ways to determine the ^{14}C content in a historical sample by measuring the activity and then by counting directly the number of ^{14}C atoms, before coming back to the question of radiocarbon variations in the atmosphere and the necessary corrections that have to be taken into account for these variations in final age determination.

Figure 5.4: The mummy of Ramesses II (1303–1213 BC) at the Egyptian Museum in Cairo http://www.quizmastertrivia.com/2016/02/todays-article-ramesses-ii.html.

5.3.2   TRADITIONAL RADIOCARBON DATING TECHNIQUES

Traditionally the number of ^{14}C atoms in the sample had to be measured by determining the β activity of the sample

$$A(^{14}\text{C}) = \lambda \cdot N(^{14}\text{C}). \tag{5.2}$$

This is not easy because ^{14}C decays by the emission of low-energy 152 keV β radiation. At such energies, the β particles (electrons) are easily absorbed into the environment before being detected. Missing electrons would clearly result in an artificial increase in the estimated age, since ^{14}C decay would have been interpreted to have occurred at a lower rate than it actually had. Special attention has therefore been given to developing methods for optimizing the efficiency of β detection. This requires chemical procedures on the sample itself, since the decay radiation from inside a solid sample is being absorbed in the sample itself. Therefore, the sample has to be converted to gaseous carbon dioxide ^{14}CO$_2$ or to methane ^{14}CH$_4$, to be used as counting-gas in a gas detector like a Geiger counter. Careful measurements on calibrated samples are necessary in order to determine the efficiency of the specific instrument.

The chemical preparation of the samples requires a thorough cleaning of the initial material samples to remove any more-recent contaminations, which would translate the results to a younger date. A 1% contamination with modern carbon of a 23,000-year-old sample would translate into a younger age of 21,700 years. A larger amount of contamination will obviously result in a substantially larger deviation. Often HCl acid is used for removing carbonates and other contaminants. Absorption of CO$_2$ gases from the air must be prevented. Finally, the sample is burned to convert it to CO$_2$ or processed to CH$_4$ for the actual counting process. Carbon dioxide is chemically easy to produce without substantial material losses, but the chemical production and purification process carries some electronegative impurities such as chlorine, which may affect the counting, when using Geiger counters.

The use of the gas-counters was for a long time the preferred technique. The gas volume typically contained a few liters at a pressure of 1–2 bar. This represents about 2 g of carbon, with a small fraction of ^{14}C. This leads to 25–30 decays per minute, for a sample of more recent age. To be sensitive to the reduced count rate from older samples, possible background in the counting detectors has to be minimized. Most of the background typically comes from cosmic radiation and the radiogenic background produced by the decay processes in the surrounding material. The background count rate in the detectors can be reduced to 1–2 events per minute by using local lead- and steel-shielding as well as anti-coincidence techniques, with cosmic ray detectors surrounding the counting system. This means that after about 3–4 half-lives, the actual β count rate from the ^{14}C decay is comparable with the background count rate, which limits analysis to about 23,000 years.

Alternatively, the initial sample material can be converted chemically into an organic solvent such as benzene ($^{14}C_6H_6$). These solvents scintillate when radiation is released within their volume. One can use the benzene as a liquid scintillator in connection with a highly efficient photomultiplier tube to measure the internal ^{14}C radioactivity of the scintillation liquid. This requires substantial chemical preparation, which involves extracting a fixed amount of carbon from the sample and dissolving or integrating it into the scintillator material in a controlled manner. Despite carefully documented chemical steps, these procedures introduce additional uncertainties into the final detection results.

There are two challenges, first, the sample must contain a sufficient amount of ^{14}C to be detectable compared with other natural radioactive material in the detector material, and the ^{14}C must also be detectable against a steady background of cosmic ray induced background signals in the detector itself. For the classical counting techniques, this level of background corresponds to the activity of about a gram of carbon material in the detector. Therefore, an appreciably higher amount of the valuable original sample has to be sacrificed to be converted into detectable counting material.

5.3.3 ^{14}C ACCELERATOR MASS SPECTROMETRY (AMS)

A major innovative development in radiocarbon dating occurred in the late 1970s and early 1980s with the development of Accelerator Mass Spectrometry (AMS). This technique allows for directly counting the number of ^{14}C atoms in a sample, rather than measuring ^{14}C decay activity. Counting $N(^{14}C)$, rather than measuring $A(^{14}C)$, improved the sensitivity of the method enormously and made it possible to analyze much smaller samples for ^{14}C content. While previously, relatively large samples (grams) had to be graphitized to be converted to scintillator or counting gas material, using the AMS method small amounts (milligrams) are sufficient to count the number of radioactive ^{14}C atoms reliably.

The AMS technique became possible after the development of the so-called tandem accelerators. These machines are based on the Van de Graaff principle of the classical electrostatic accelerator. Negative ions are produced in a so-called sputter source, where the ion source material is bombarded by a low-energy cesium beam; the atoms sputter off the pellet and pick up negative charge from the cesium atoms, to be extracted out of the ion source area. An injection magnet system is used for a first e/m analysis of the extracted ions before they are accelerated to a high-voltage terminal at a terminal voltage V. At the terminal, the negative ions are stripped to a positive charge state distribution by running through a so-called stripper foil or stripper gas and are subsequently accelerated through the second acceleration stage of the tandem, reaching energies of $E = (1 + q) \cdot V$ with q being the electrical charge of the positive stripped ion. An analyzing magnet serves to select one of the charge states of the accelerated ion for the final experiment.

The basis of the development of AMS techniques was the recognition that it was not possible to produce negative ^{14}N ions. This meant that no negative ions of mass $A = 14$ could be produced in the ion sources. Even spurious ^{14}N components would have represented a major background for the counting of single ^{14}C ions and would have made AMS impossible. However, while there are no ^{14}N contaminations in the negative ^{14}C$^-$ beam, negatively charged molecular impurities such as ^{12}CH$_2{}^-$ or ^{13}CH$^-$ could well contaminate the beam, leading to a substantial $A = 14$ background count rate as demonstrated in Figure 5.5. These contaminations are removed by the stripper foil of the tandem, thereby breaking up the molecules into their single-particle fragments that will be easily removed by the analyzing magnet.

The goal in AMS systems is not the absolute counting of the ^{14}C ions released from the sample but, rather, the determination of the ^{12}C/^{14}C or sometimes also the ^{14}C/^{13}C ratios, which have developed from the atmospheric ratio at the death of the original biological system, through the decay of the ^{14}C isotope. Both ^{13}C and ^{12}C are stable isotopes and their role in the abundance ratio does not change with time. For this reason, not only ^{14}C but also the stable carbon components have to be analyzed and counted in the AMS system. The actual counting typically uses ion chambers in which the carbon particles are being stopped. In many cases, the energy loss of the particles is also being measured as a further way to identify and reduce background counts, from randomly scattered particles, in the detector system.

There are different options for positioning the ion counters for the various carbon isotopes. This positioning depends on the specific layout of the system. Figure 5.6 shows a typical AMS system as it was developed at Lawrence Livermore Laboratory in the 1990s. ^{12}C ions are the dominant species in the beam, being ejected from the ion source, and are therefore easily separated by the injection magnet, which serves as a simple mass analyzer for these ions; all heavier beam components are injected into the tandem, where the molecular fragments are being broken up into the light ion constituents. The ^{13}C counting station is located after the analyzing magnet. Further cleaning of the beam happens in an electrostatic analyzer or a Wien filter system that serves as velocity analyzer, removing randomly scattered particles with differ-

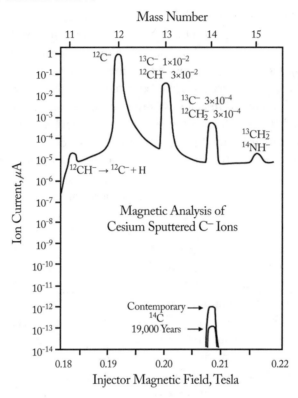

Figure 5.5: Mass analysis of a carbon sample after the first injection magnet past the ion sputter source. The ion composition is primarily characterized by molecular impurities in the beam; the anticipated radiocarbon component of both modern and 19,000-year-old material is many orders of magnitude smaller than the molecular background, which needs to be removed by the tandem accelerator system https://www3.nd.edu/nsl/Lectures/phys10262/art-chap3-5.pdf.

ent velocities from the beam before the final ^{14}C counting system. More modern AMS systems such as VERA at the University of Vienna, Austria or CIRCE at the University of Caserta, Italy, count both ^{12}C and ^{13}C after the tandem analyzing magnet, to remove possible light ion $A = 12$ molecular components from the beam.

Today, AMS is a standardized, widely used technique at many laboratories worldwide. New developments have allowed for optimization of the AMS method and for the use of smaller and smaller accelerators. While the first AMS systems were based on massive tandem systems with terminal voltages of more than 10 MV, more modern systems are substantially smaller; terminal voltage has been reduced to 3 MV. More recent developments have aimed at AMS systems of table-top size, which are now commercially offered by many accelerator companies

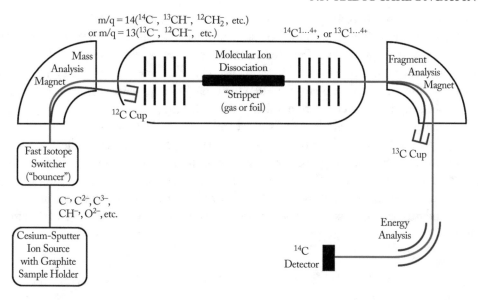

Figure 5.6: Typical AMS system with tandem accelerator at Lawrence Livermore National Laboratory. The ^{12}C intensity is being measured before the tandem, as the strongest component in the sputtered mass distribution. ^{13}C is being measured after the molecular components have been removed, by break up in the stripper foil and rejection in the magnetic field of the analyzing magnet. After this, a final electrostatic or, often, Wien filter station helps to remove further multiple scattered particles that could be a background for the final ^{14}C counting.

since there is an increasing demand for AMS systems beyond the realm of cultural heritage studies.

The experimental determination of the ^{12}C/^{14}C isotope ratio in the sample yields a ratio that is lower than the atmospheric ratio at the present time. Assuming that the atmospheric ratio has been constant, the radiocarbon age of the sample can be directly determined using the formalism described below:

$$t = \frac{1}{\lambda} \ln\left(\frac{^{14}C(t=0)}{^{14}C(t)}\right) = 8284 \cdot \ln\left(\frac{^{14}C(t=0)}{^{14}C(t)}\right)$$

$$t = 18500 \cdot \lg\left(\frac{^{14}C/^{12}C(t=0)}{^{14}C/^{12}C(t)}\right).$$

(5.3)

Assuming the current ^{12}C/^{14}C abundance ratio of $1.3 \cdot 10^{-12}$ at $t = 0$, (the time of death), the age can be directly calculated from the measured ^{12}C/^{14}C abundance ratio. However, the atmospheric ^{12}C/^{14}C abundance ratio is not constant and has dramatically changed with time, as the tree ring analysis has demonstrated. To monitor this requires a reliable reference and a normalization system is necessary, since the ^{14}C abundance in the atmosphere and in biological materials is usually not given in terms of absolute numbers. During the nearly 70 years of radiocarbon dating, such a system of references and reference frames have been developed that are still based in part on early measurements of ^{14}C activity and not on the counting of ^{14}C atoms. A number of different ways have emerged in which the ^{14}C abundance can be expressed and it is always a challenge to convert these numbers into each other.

Frequently, also, the abundance of the second stable carbon isotope ^{13}C is involved as a reference point, since it allows for monitoring fractionation processes that may lead to changes in abundance ratios, having nothing to do with the age of the sample. To correct for these changes, the deviations are expressed as differences to a standard value, using the delta (δ) notation as the deviation in parts per thousand of the ratio $R = {}^{13}C/{}^{12}C$ to the ratio measured in an universally accepted standard, which is based on a belemnite fossil found in the Pee-Dee formation in South Carolina.

5.3.4   GLOBAL VARIATIONS IN ATMOSPHERIC RADIOCARBON ABUNDANCE

As pointed out earlier, the radiocarbon content in the atmosphere is not constant but depends critically on the atmospheric composition. This has multiple reasons and requires careful analysis as well as correction of original data. Reasons for these content differences include variation in the intensity of cosmic rays, the level of fossil fuel in use, the nuclear bomb test program, chemical fractionation in the material to be analyzed, to name only the most important considerations. To understand the impact of these effects, careful calibration measurements and the use of well-known standard material is imperative.

Cosmic ray intensity depends on solar activity and the strength and direction of the earth's magnetic field as discussed earlier. Continuous exposure of the earth's atmosphere to cosmic rays facilitates the production of ^{14}C. So, ^{14}C is directly associated with the level of solar activity and the strength and direction of the earth's magnetic field. This occurs due to the abundance of ^{14}N in the atmosphere; although this target nucleus is nearly constant, a high intensity of the cosmic ray flux will produce a higher neutron flux, for the ^{14}N(n,p)^{14}C reaction, thereby increasing ^{14}C production. This can be clearly observed in the analysis of old wines, where the ^{14}C content varies with the 11-year cycle of solar activity, as shown in Figure 5.7. Figure 5.7 also demonstrates a second effect, the Suess effect, which is reflected in the gradual decline of the ^{14}C abundance in wine material.

The fossil fuel effect was first predicted by Hans Suess, one of the pioneers in isotope analysis and radiocarbon dating; it is, therefore, often called the Suess effect. This is a human-made effect, which is due to the increasing use of fossil fuel associated with industrialization and traffic. Fossil fuel is millions of years old and the original radiocarbon content has long since decayed. With the use of fossil fuels such as coal, oil, and gas, the carbon content is released as CO_2 in the burning process. However, the ^{14}C component in the released carbon dioxide is

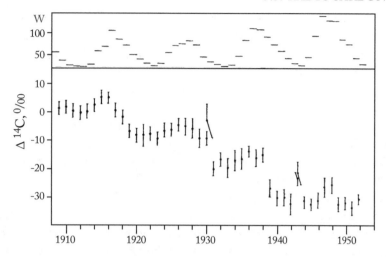

Figure 5.7: Sunspot activity and the relative ^{14}C content in Georgian wine samples harvested during the period of 1909–1952. The figure demonstrates the direct correlation between solar activity and ^{14}C deposition in the atmosphere. Figure plotted from the data presented in Burchuladze, A., Pagava, S., Povinec, P. et al. Radiocarbon variations with the 11-year solar cycle during the last century. *Nature*, 287:320–322, 1980.

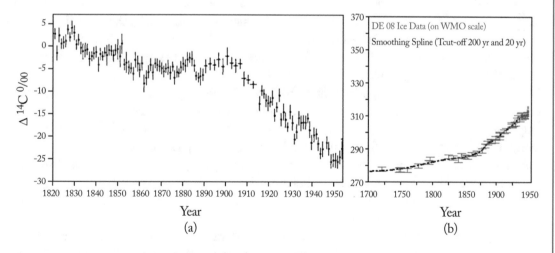

Figure 5.8: Radiocarbon content of the tree rings of Douglas fir (Pseudotsuga menziesii) on the west coast of the USA in the industrialized period. The clear reduction of 14C begins ca. 1900 (a) (Stuiver and Quay 1981). The increase of fossil fuel CO2 emission (in ppm) into the atmosphere as measured from ice-core data (b). The increase in fossil fuel emission between 1900 and 1950 is well reflected in the relative decline of ^{14}C over the same period. Figures plotted from the data presented in M. Stuiver and P. D. Quay. Atmospheric ^{14}C changes resulting from fossil fuel CO_2 release and cosmic ray flux variability, *Earth and Planetary Science Letters*, 53:349–362, 1981.

nil, causing a relative increase in $^{12}CO_2$ content in the atmosphere and an overall decrease of the ^{14}C content. This is clearly observed in Figure 5.8, which shows the development of ^{14}C in the tree rings of the Douglas fir over the entire period of fossil fuel based industrialization from 1820–1950. While the first decades are characterized by variations that can be correlated with the 11-year solar cycle, it can clearly be observed that the decline of ^{14}C starts with the turn of the 19th century.

This is, of course, a critical component in the radiocarbon dating process since the radiocarbon calibration curve is directly affected, in particular when analyzing samples from the industrial period. More recently, however, a much stronger effect is due to the nuclear weapons test program that took place in the second half of the 20th century.

The nuclear weapons test program has caused an additional anthropogenic effect in recent history, which has resulted in dramatic variations of the ^{14}C content in the atmosphere. This is reflected by a dramatic increase of ^{14}C in the atmosphere with the beginning of the nuclear test program of the United States, followed by the test programs of the Soviet Union and other nuclear powers. In the period between 1950 and 1970, more than 2,000 atmospheric tests were performed, and each explosion released a substantial flux of neutrons into the atmosphere. The initial tests focused on fission bombs, which rely on neutron induced fission of either ^{235}U or ^{239}Pu fossil material into smaller fission fragments. Each fission process releases, on average, three neutrons, and the fission products come in a double hump distribution (as a function of their mass number), which reaches far into the range of *very neutron-rich nuclei*. The neutron flux released in such a traditional *plutonium of uranium fission bomb* test is enormous; there is the release of neutrons that are directly released in the fission, but there is also the so-called delayed neutron emission, which is released in the decay process of very neutron-rich fission products. Figure 5.9 shows the neutron flux emitted as a function of time by a typical fission bomb such as the Nagasaki bomb of 1945.

A fusion bomb—traditionally called a hydrogen bomb—is based on the principle that the fusion of *light neutron-rich* hydrogen isotopes to heavier nuclei will release energy. Fusion can only occur at high temperatures in order to overcome the Coulomb barrier between two positively charged particles. Therefore, fusion bombs are triggered by a small fission device. The most common fuel for fusion bombs is a mix of neutron-rich hydrogen isotopes—such as deuterons ^2H and tritons ^3H—which fusion via the reaction $^3H(d,n)^4He$ to helium, releasing neutrons in the energy range of 14 MeV. Again, the neutrons rapidly scatter down into the lower energy range.

Neutrons from both fission and fusion events have the same impact as cosmic ray induced neutrons, after losing their initial high energy through multiple scattering events, in fractions of a second, before they interact with the ^{14}N content in the atmosphere through the $^{14}N(n,p)^{14}C$ reaction, enhancing the ^{14}C content in the atmosphere.

This process is impressively confirmed by the so-called bomb peak, the dramatic increase in ^{14}C in the atmosphere throughout the atmospheric nuclear bomb test program from 1945–1969

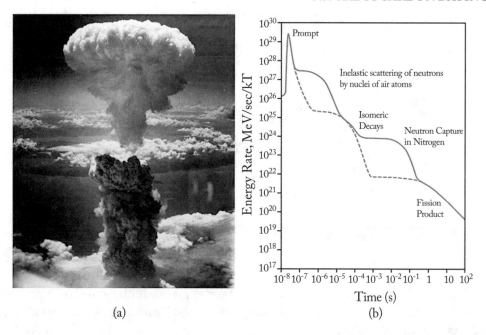

(a) (b)

Figure 5.9: Mushroom cloud of the plutonium bomb dropped in August 1945 on the Japanese city of Nagasaki, releasing a high flux of neutrons into the atmosphere (a) https://www. atomicheritage.org/article/anniversaries-atomic-bombings. The neutron flux of a fission bomb as a function of time (b). The high flux in the very early moments is caused by the prompt neutrons released in the actual fission process, followed by the delayed neutrons, which typically have lower energies. All neutrons are being slowed down by multiple scattering processes. With decreasing neutron velocity, the probability for the $^{14}N(n,p)^{14}C$ reaction increases, through which a substantial number of neutrons are absorbed as indicated https://www.dtra.mil/Portals/61/ Documents/NTPR/4-Rad_Exp_Rpts/36_The_Effects_of_Nuclear_Weapons.pdf.

(Figure 5.10). This peak appears in observations in the northern as well as southern hemisphere, verifying claims that high altitude winds distribute the freshly produced ^{14}C rapidly over the entire stratosphere and troposphere.

The half-life of ^{14}C is 5,730 years, yet the decline of atmospheric radiocarbon is much faster. In approximately 10 years after the highest ^{14}C abundance in 1965, the ^{14}C had dropped to 50% of its peak value. This is due to the chemical carbon cycle, in which carbon couples with atmospheric oxygen of CO_2, which is absorbed by ocean water and plant material. The bomb peak can, therefore, be used as a signature for the exchange of atmospheric CO_2 and the terrestrial ecosystems, such as in oceans and bio-mass that have absorbed the long-lived ^{14}C.

The bomb peak is a unique feature and literally re-set the clock for radiocarbon dating. Elevated ^{14}C content in materials points to the time of its origin, during the bomb testing era.

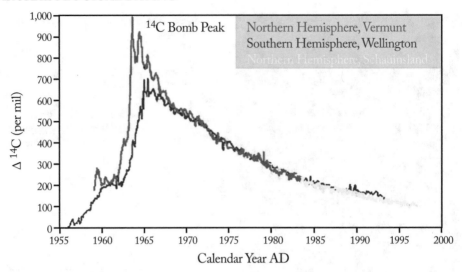

Figure 5.10: The ^{14}C bomb peak in atmospheric measurements. Initially the bomb peak on the southern hemisphere is stronger due to the early tests in the Pacific but with the opening of the Nevada test site and the Soviet bomb test program, the bomb peak in the southern hemisphere is delayed by about one year due to the atmospheric distribution mechanism of high altitude winds (https://www.esrl.noaa.gov/~gmd/outreach/isotopes/bombspike.html).

This emerged as an interesting tool in the analysis of anthropological samples and in the identification of recent forgeries. However, it is also an important tool for testing medical and biological samples, such as identifying the age of blood or brain cells, leading to a better understanding of cell regeneration mechanisms. The later use is being reflected in the increase of small AMS facilities used by pharmaceutical companies for their specific research needs.

Atmospheric calibration is necessary to account for the variations in the amount of ^{14}C in our atmosphere, which we have just discussed. Determining the actual amount of ^{14}C in the atmosphere at the time of the origin of the sample is an absolute necessity for determining the age of the sample from the measured ^{14}C/^{12}C ratio in AMS studies. This calibration is one of the primary sources of uncertainty in age determination and is typically based on dendrochronology methods, where the ^{12}C/^{14}C ratios are determined for tree ring sections of well-known age (Figure 5.11).

Chemical fractionation is a phenomenon that causes changes in the abundance ratios between different isotopes due to chemical or physical means over extended periods of time. While the tabulated isotopic ratios represent average values within the solar system, substantial deviations can occur in specific samples, depending on the sample history. One example, discussed in Chapter 4, is the ratio of ^{18}O to ^{16}O isotopes, reflected in the human tooth, hair, and bone material. In rainy coastal regions, the ^{18}O content is enhanced, since water molecules

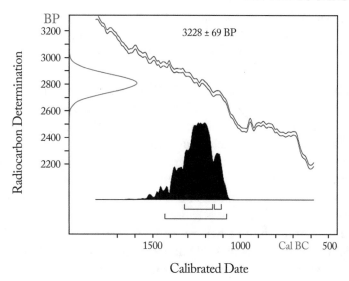

Figure 5.11: Correction of the radiocarbon age of a specific sample, converted to the real age by using the section of the calibration curve for that specific age range https://c14.arch.ox.ac.uk/calibration.html.

H_2O including ^{18}O fall more easily with rain due to their heavier weight. The clouds become gradually depleted in ^{18}O as they drift inland. Therefore, rainwater far from the coast contains a lower ^{18}O content, as can be easily measured with high-precision mass analyzers [1].

For the ^{13}C to ^{12}C ratio, the fractioning mechanisms are based on the photosynthesis of plants, which naturally enhances the lighter carbon isotopes so that the carbon in plants is depleted of heavier carbon isotopes such as ^{13}C and also ^{14}C [2]. After photosynthesis, the isotope ^{13}C is depleted by 1.8% in comparison to its natural ratios in the atmosphere.

Fractionation is expressed in terms of $\delta^{13}C$, which is a measure (in parts of 1000 ppm) of the deviation of the isotopic ratio $^{13}C/^{12}C$ from a standard material (PDB belemnitella americana):

$$\delta^{13}C \equiv 1000 \cdot \frac{\left[\left|\frac{^{13}C}{^{12}C}\right|_{sm} - \left|\frac{^{13}C}{^{12}C}\right|_{st}\right]}{\left|\frac{^{13}C}{^{12}C}\right|_{st}} = 1000 \cdot \left[\frac{\left|\frac{^{13}C}{^{12}C}\right|_{sm}}{\left|\frac{^{13}C}{^{12}C}\right|_{st}} - 1\right]. \tag{5.4}$$

Typical $\delta^{13}C$ varies between +2 ppm to −27 ppm and needs to be determined for the material to be dated. A negative value $\delta^{13}C$ means that the sample is isotopically lighter than the standard probe. The standard is the fossil belemnite from the Pee Dee formation in South Carolina, PDB, $(^{13}C/^{12}C)_{PDB} = 0.0112372$. A positive value means that the sample is enriched in the heavier isotope components.

If one has a material with the average $(^{13}C/^{12}C)_{sm}$ ratio of 0.01112235, the value of the fractionation parameter would be $\delta^{13}C = -10.221$. A reduction in the $^{13}C/^{12}C$ ratio by 1.8% would shift this value to $\delta^{13}C = -28.036915$. This is a rather dramatic change that has to be accounted for.

Isotopic fractionation will also affect the $^{14}C/^{12}C$ ratio, which is the critical parameter for radiocarbon dating. Fractionation will be much more severe since the physical causes underlying the phenomenon will be much more pronounced due to the larger size and the larger mass of the ^{14}C isotope. Its mass is approximately double that of the measured $^{13}C/^{12}C$ ratio, meaning that one would expect a reduction of approximately 3.6%. This affects the radiocarbon dating of biological materials since dating depends on the kind of plant materials involved or the kind of food intake prior to the time of death, of creature or man whose leftovers have been analyzed. In all cases a reduced ^{14}C content is expected in comparison with the ^{14}C content in the atmosphere at the time of death. This requires another correction, since a reduced ^{14}C carbon content in the original sample material would indicate a higher age for the sample at the time of measurement. These corrections are challenging since they require at least some knowledge or reliable interpretation of the origin and later history of the sample, during which other possible fractionation events might have occurred [3].

The fractionation of ^{14}C is defined in the same terms as for ^{13}C, assuming that $\delta^{14}C = 2 \cdot \delta^{13}C$. Following this, the age correction due to fractionation can be approximated by a simple empirically derived formula for material with a certain fractionation of $\delta^{13}C$ in units ‰:

$$t_{corr} - t_{uncorr} \approx 16 \cdot \left(\delta^{13}C + 25\right) \text{[y]}. \tag{5.5}$$

In view of these analytical complexities, strict international regulations have been developed to compare and normalize the results from AMS radiocarbon analysis at different laboratories. These rules suggest how the analysis has to be performed and which corrections need to be adapted, as well as which standard samples for calibration need to be used.

1. The standard is a sample of oxalic acid by the U.S. National Bureau of Standards, which provides the content of ^{14}C for the year of 1950, prior to the nuclear bomb tests. The concentration is adjusted in such a manner that it compares to the atmospheric concentration of 1850, to account for the Suess effect.

2. The use of the ^{14}C half-life of 5,568 years provided by Libby. A new half-life of 5,730 years has been established later, which means that the standard radiocarbon age needs to be corrected by the 3% increase.

3. A correction for fractionation effects through a measurement of the $^{13}C/^{12}C$ ratio and the extracted value is normalized to $\delta^{13}C = -25$ ppm.

4. Age is given in years BP (before present), with the present being 1950 AD.

Figure 5.12: Ötzi at the time of his discovery in a melting ice water pool in the Ötztal https://www.donsmaps.com/otzi.html.

5.3.5   EXAMPLES OF AMS APPLICATIONS

AMS radiocarbon dating has become a standard technology. AMS laboratories exist worldwide, particularly in Europe where the analysis of cultural heritage material has emerged as a major effort (https://www.edp-open.org/images/stories/books/fulldl-/Nuclear-physics-for-cultural-heritage.pdf). AMS analysis is now routine and has even become standard procedure in medical and pharmaceutical applications, using the bomb peak as point zero for the radiocarbon clock.

In terms of cultural heritage materials, numerous examples of radiocarbon dating are published in the respective literature. The purpose of most of the measurements is to complement the information base for research and excavation results of the anthropology and archeology community, while other results gain high visibility because of the religious and sociological impact.

A high visibility example has been the age determination of the so-called ice-man or Ötzi for short, referring to the place of his discovery. This mummified body of a man was found, in September 1991, in a puddle of melted ice water at high altitude in the Ötztal Alps, right at the border of Austria and Italy (https://en.wikipedia.org/wiki/Ötzi). Detailed medical analysis provided a wealth of data about his health status, his food consumption, his origin in the southern range of the alpine range on his way north, and also that Ötzi was killed by an arrow shot. The body was later covered by snowfall, which turned to glacial ice and therefore remained in a well-preserved state until released by melting, thousands of years later (see Figure 5.12).

Ötzi was a unique discovery. The body material and the material of iceman's equipment were immediately radiocarbon dated. AMS method was chosen because it required only miniscule body samples ($\approx$ 1 mg). AMS measurements were performed independently at five different AMS accelerator labs, the ETH, Zürich, Switzerland [4], the University of Oxford, UK [5], and Gif-sur-Yvette, France, University of Uppsala, Sweden, and VERA, University of Wien, Austria [6].

The combined un-calibrated radiocarbon age is 4550 $\pm$ 19 y BP (BP: before present 1950) 2596 $\pm$ 17 BC. Calibration against the tree ring curve is necessary to account for variations in the atmospheric ^{14}C concentration. The intersection of the un-calibrated radiocarbon age with the tree ring calibration will show the calibrated age of the sample. The detail of the calibration over the period of 3500–3000 BC indicates three possible intersections due to the statistical uncertainty of $\pm$ 19 years (Figure 5.13). The calibration shows that Ötzi is about 650 years older than the radiocarbon date. However, the intersections suggest three options for the time range in which Ötzi was murdered, 3350–3300 BC with 56%, 3210–3160 BC with 36%, and 3140–3160 BC with 8% probability, taking into account statistical uncertainties of the various measurements and uncertainties associated with the tree ring calibration curve. This age range was confirmed by the radiocarbon dating of equipment and other human-made materials found near the body [7]. Other human-made materials that were found in larger distances were radiocarbon dated as well. The materials showed a considerable age range and indicate that the pass was used as a transition route between the southern and northern slopes of the Alps at significantly earlier and later times, 2000–600 BC.

Another high visibility study, which caused considerable controversy, is the AMS dating of the shroud of Turin, which was independently done by three AMS laboratories in Arizona, Oxford, and Zürich [8]. The Shroud was supposedly used to wrap Christ's body. It bears detailed front and back images of a man who appears to have suffered whipping and crucifixion. It was one of the numerous relics that had been brought back by crusaders returning home from the Holy Lands and was first displayed at Lirey in France from 1357–1418 and was then exhibited at Saint Hippolyte-sur-Doubs from 1418–1452. The Shroud subsequently passed into the hands of the Dukes of Savoy and was kept in the chapel of Chambery Castle from 1454–1578. During that period, it was also displayed as a relic of particular importance at many other places.

The shroud was declared authentic by Pope Julius II in 1506, independent of the existence of several other "medieval" shrouds at Besançan, Cadouin, Champiègne, Xabregas, which were claimed to be of linen provided by Joseph of Arimathea for wrapping the body of Jesus after the crucifixion.

In 1532, a fire broke out in the chapel of Chambery, partly damaging the Shroud. In 1578 the shroud was finally brought to Turin, where, in 1694, it was placed in the royal chapel of Turin Cathedral in a specially designed shrine. In 1997, fire broke out in the dome of the Cathedral in Turin. The Shroud was not damaged in any way, but may have been exposed to smoke and other exhausts.

The samples for the AMS analysis were selected by representatives of the church. The procedure is described in detail in the paper: *The shroud was separated from the backing cloth along*

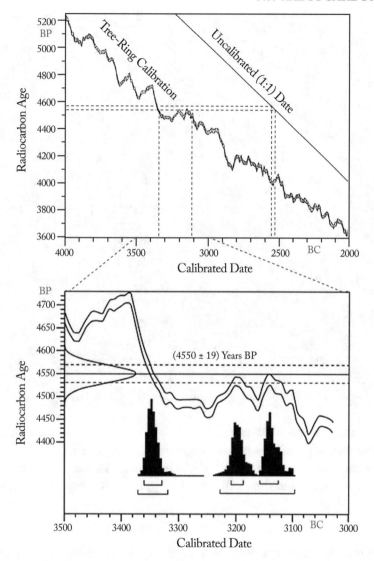

Figure 5.13: The determination of the age of the Iceman from ^{14}C measurements at the AMS laboratories. The combined radiocarbon age from these measurements is 4546 ± 17 years BP (Before Present = 1950 AD). The error is the 68.2% (1σ) confidence value. The uncalibrated age is translated into a calibrated age using a tree-ring calibration curve. (a) Calibration curve from 4,000–2,000 BC. The straight line at 45° indicates a 1:1 transformation of the radiocarbon age into an uncalibrated calendar date. The intersection of the radiocarbon age with this line and the tree-ring calibration curve shows that the calibrated date is approximately 650 years older. (b) The enlarged "wiggly" section of the calibration curve leads to 3 different solutions for the calendar date, spanning 250 years. Reprinted with permission from [7]. Copyright 2000 Elsevier.

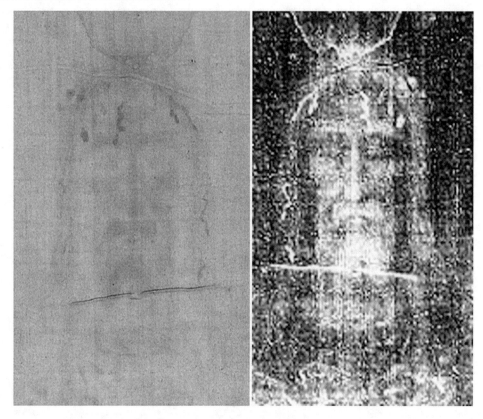

Figure 5.14: On the left, is a photographic "negative image," as it appears on the cloth itself. When photographed, a "positive" image emerges as the film is being developed. Claim is that the negative resulted from chemical reactions between ammoniac evaporations (or bacteria) from a tortured body (of Christ) and the cloth material https://www.shroudofturin.com/Resources/CRTSUM.pdf.

its bottom left-hand edge and a strip (∼ 10 mm × 70 mm) was cut from just above the place where a sample was previously removed in 1973 for examination. The strip came from a single site on the main body of the shroud away from any patches or charred areas. Three samples, each ∼ 50 mg in weight, were prepared from this strip. The samples were then taken to the adjacent Sala Capitolare where they were wrapped in aluminum foil and subsequently sealed inside numbered stainless-steel containers. [...] Samples weighing 50 mg from two of the three controls were similarly packaged. The three containers containing the shroud (to be referred to as sample 1) and two control samples (samples 2 and 3) were then handed to representatives of each of the three laboratories together with a sample of the third control (sample 4), which was in the form of threads. [...] The laboratories were not told which container held the shroud sample.

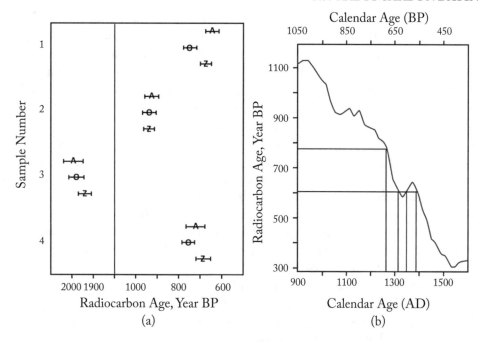

Figure 5.15: Mean radiocarbon dates, with ± 1 u errors, of the Shroud of Turin and control samples (a), as supplied by the three laboratories (A, Arizona; O, Oxford; Z, Zurich). The shroud is sample 1, and the three controls are samples 2–4. Note the break in age scale. Ages are given in year BP (years before 1950). The age of the shroud is obtained as AD 1260–1390, with at least 95% confidence. Calibration of the overall mean radiocarbon date for sample 1 (the Shroud of Turin) using the *intercept* method to accommodate for the natural variations in atmospheric ^{14}C (b). The uncertainty in the calibration curve has been combined with the error in the mean radiocarbon date, giving the 95% confidence limits on the radiocarbon scale. The stippled areas show how the 95% confidence limits are transformed from the radiocarbon to the calendar date. Figure plotted from the data presented in [8].

The test samples were linen taken from Nubian tomb, which was 950 years old, linen from an Egyptian mummy wrapping, 1,900 years old, and threads from a cape of St. Louis d'Anjou, 700 years old.

The AMS experiments all independently confirmed the dates of the three test samples within the given experimental uncertainties. They also independently determined the age of the shroud sample to be from the period of 1262–1312 or 1353–1384 AD with a 95% confidence level. This would indicate that the shroud was a medieval forgery, presumably sold for good money to a medieval crusader on his way home.

It seemed to be a convincing measurement in the eyes of the scientists. Nevertheless, it met with strong criticisms from the religious community, which sought to identify reasons for the strong deviation between the officially stated age of the shroud and the radiocarbon results. This criticism is based on a number of arguments including the fact that the shroud was exposed to high temperatures above 300°C as indicated by burn marks and molten silver spots. This could have been due to the exposure to fire and smoke during the fire in the chapel of Chambray Castle in 1532, possibly causing fractionation between ^{14}C and ^{12}C, causing an artificial reduction of age. In addition or alternatively, intense incense and candle burning in front of the shroud would have added modern age contaminants with higher ^{14}C abundance, which again would reduce the radiocarbon age. Yet, given the large difference in the claimed age and the radiocarbon result, the selected portion of the shroud would have had to experience a doubling of the $^{14}C/^{12}C$ ratio in its material, to double the age. It was stated that *a contaminated 1.8 kg first-century shroud would weigh more than 5 kg with more than 3 kg of contaminant to yield a 1355 date* [9]. These arguments underline the importance of understanding the history of the biological samples in order to ensure a reliable age determination.

Another source of concern was about the methodology of the systematic analysis procedure employed by the team, in particular on the comparison of the results and the associated statistical relevance. To avoid going into the wealth of scientific and semi-scientific publications and statements, these arguments are summarized and repeated in a paper re-analyzing the original samples using modern statistical tools [10]. The authors claim that there are misgivings regarding the 95% reliability statement on the concluded age of the shroud in the original analysis. This does not mean that the previous conclusions on the age of the shroud are wrong but, rather, that the certainty of the claim is diminished. The authors suggest a re-measurement with an improved statistical analysis method. This is certainly a sensible suggestion, however, it will be challenging to come to a conclusion, *among all this feverish speculation, for calm science to prevail, especially because no further testing of the shroud—venerated last year by Pope Francis as an "object of devotion"—has so far been authorized* [11].

5.4   THERMOLUMINESCENCE

Thermoluminescence (TL) is a widely used method for the age determination of crystalline cultural heritage samples. While the radiocarbon method is confined to organic materials and its carbon component only, TL is confined to inorganic materials such as stone and pottery, which are either of crystalline structure or contain spurious crystalline contents. This method can only be used for insulating material and not for metallic artifacts (see Figure 5.16).

Thermoluminescence dating is a technique that is based on the analysis of light release when heating such materials. TL-dating is used in mineralogy and geology but is also increasingly being applied for dating of anthropological and archaeological samples, where it has emerged as an important tool for the quick and inexpensive testing of cultural heritage materials in the identification of forgeries. In archaeology, TL is mainly used for pottery analysis, while

Figure 5.16: A Mesoamerican pottery figure and a crystalline stone exhibiting a TL glow after being heated to more than 200°C.

in anthropology, the primary use of TL is the dating of flint stone as early tool material for humankind.

The typical phenomenon of TL is the release of light when a sample is heated above 200°C. Light emission in the blue range is observed up to 400°C. At higher temperature, the material emits a red glow (Figure 5.17). In second heating, no blue light emission is observed; only the red glow curve remains at the higher temperature range (Figure 5.17).

Thermoluminescence originates from a temperature-induced release of energy stored in the lattice structure of the crystal, following long-term internal and external exposure to nuclear radiation from long-lived radioactive elements, such as ^{238}U, ^{232}Th, and ^{40}K, which occur naturally in nature. With the radioactive decay of these elements, which pumps energy into the crystalline structure of the stone material, TL accumulates in the material over time, depending on radiation (and light) exposure. The TL is released by heating. The dating clock starts with the initial firing of the material when originally accumulated TL was driven out (see Figure 5.18).

The amount of accumulated TL is proportional to the age of the sample and inversely proportional to the nuclear radiation exposure of the sample, which can be expressed in the simple relation:

$$\text{Age} = \frac{\text{Archaeological TL}}{(\text{Annual Dose}) \cdot (\text{TL/unit dose})}. \tag{5.6}$$

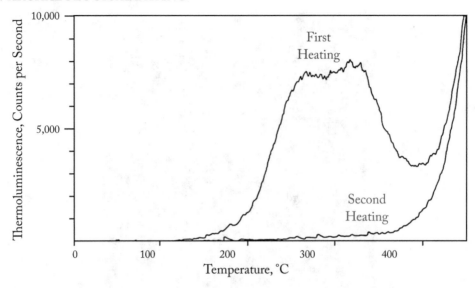

Figure 5.17: A typical glow curve observed in a TL analysis. The light emission during the first heating of the object is in the blue range and the intensity corresponds to the age of the object. Further heating generates a red glow. This red glow is observed in a second heating of the object, but no blue light emission is observed.

Dose D is defined in Section 1.7.4 as the amount of energy E deposited by radiation exposure into a sample of mass m within one year. D depends on the content of the radioactive nuclei in the sample material and on exposure to external radioactivity:

$$\frac{D}{t} = \frac{E}{m \cdot t}.$$

(5.7)

The international standard unit is the Gray (Gy), with 1 Gy = 1 Joule/kg (energy/mass). In the United States, the old unit rad is frequently still being used, with the following conversion to Gy: 1 rad = 1 centigray = 10 milligrays (1 rad = 1 cGy = 10 mGy). The nominal background radiation absorbed dose is 100 mrad/year = 1 mGy/yr.

The capacity of materials to store TL is in the material-specific parameter TL/unit dose. This simply describes the probability that a material will develop TL under radiation exposure and this has to be determined by independent analysis, typically by exposure to well defined radioactive sources. For metals, TL/unit dose = 0, meaning it cannot store any TL. Radiation exposure can be due to internal radiation, from the radioactive elements embedded in the material, but also from external irradiation, such as sunlight. The latter should be shielded after the excavation of an object because sunlight can generate additional TL, which will falsify the age.

The experimental set-up for TL measurements is straightforward, as shown in Figure 5.19. The observed TL intensity can be easily translated into the age of the sample, as the following ex-

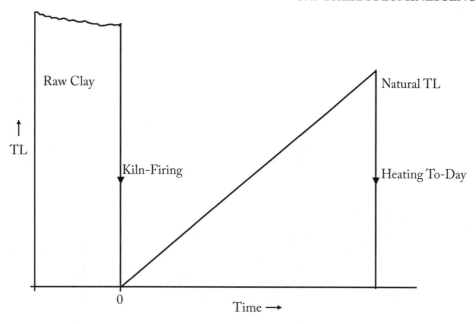

Figure 5.18: A schematic drawing showing the correlation between time and the build-up of TL in the material after firing. The build-up of TL depends on the material, literally upon how much natural radioactivity is contained in it and on its characteristic crystalline structure. The initial energy released by the nuclear decay is stored in various excitation modes of the crystal and is only released when heated.

ample demonstrates. The TL is measured first with a calibrated source to determine the TL/unit dose and then the sample itself is heated to release its TL.

The measurement of the TL with a calibrated piece of material gives 52,000 TL-cts/s after exposure to 0.1 Gy radiation. This gives the characteristic ratio TL/D = 52,0000 TL-cts/(s · Gy). The TL curve, in Figure 5.20 shows 21,000 TL-counts/s at an average annual dose of 1 mGy/y. The age can then easily be calculated using the equation given below:

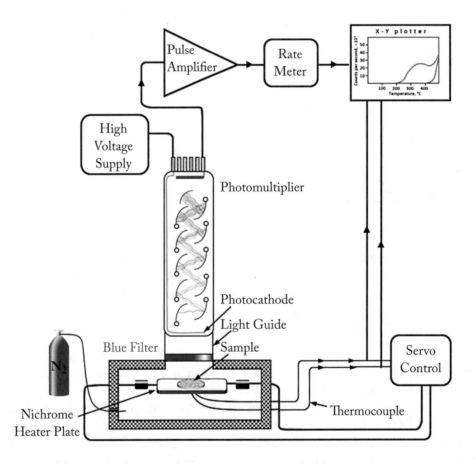

Figure 5.19: A schematic drawing showing the classical experimental set-up for a TL measurement. The sample is heated and a photocell or scintillator catches the filtered blue light. The light is converted into free electrons by the photo-electric effect and the number of electrons is multiplied in the photomultiplier to generate a signal, which is further amplified before being digitized and displayed as a function of heat.

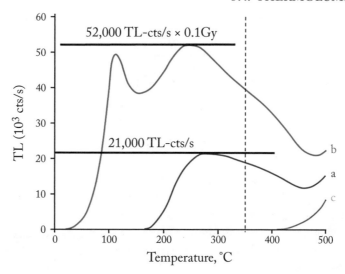

Figure 5.20: The figure shows the TL light curve of a sample, which was exposed to a well defined dose D from a radioactive source. The measured TL yield is the TL/D ratio. This ratio can be used to determine the age of the second, unknown sample from its specific TL.

$$\text{Age} = \frac{21{,}000[\text{s}^{-1}]}{1 \cdot 10^{-3}[\text{Gy/y}] \cdot 520{,}000[\text{s}^{-1} \cdot \text{Gy}^{-1}]} \tag{5.8}$$

$$\text{Age} = 4038 \text{ y.}$$

A more reliable way to determine the age is the so-called plateau test, which takes the ratio of the calibrated TL curve and the TL curve of the sample. The plateau method—initially developed as an independent test—provides a more reliable method since, with this relative approach, the systematic uncertainties in the absolute dose rates cancel out:

$$\text{Age} = \frac{\text{paleodose}}{\text{annual dose}}. \tag{5.9}$$

The paleodose N, as reflected in the emitted thermoluminescence, is the natural dose received by a sample since its last heating (production); this can be determined from the plateau ratio R between the paleodose and the paleodose plus the source related dose from the sample irradiation. The annual dose D is material dependent and needs to be measured independently.

$$R = \frac{N}{N + N_\beta}$$
$$R \cdot (N + N_\beta) = N$$
$$R \cdot N_\beta = N(1 - R) \qquad (5.10)$$
$$N = \frac{N_\beta}{(R^{-1} - 1)}.$$

If $R = 0.5(R^{-1} = 2)$ the sample has received the same dose during the source irradiation as naturally occurred over its lifetime. If $R = 0.33(R^{-1} = 3)$, the sample has received twice the natural lifetime dose during the source irradiation.

In the following, we consider the thermoluminescence measurement of a piece of ancient pottery using the plateau method. Taking the example, shown in Figure 5.21, the paleodose is calculated from the plateau ratio:

$$R = 0.47; \quad N = \frac{N_\beta}{(1/0.47 - 1)} = 0.886 \cdot N_\beta \qquad (5.11)$$
$$\text{with} \quad N_\beta = 10 \text{ Gy}; \quad \text{Paleodose:} \quad N = 8.86 \text{ Gy}.$$

With an exposure of $N_\beta = 10$ Gy, a paleodose of 8.86 Gy is being calculated. The annual dose D for typical pottery samples and soil conditions can be estimated on the basis of the exposure to decay radiation from ^{40}K and the natural decay chains of ^{232}Th and ^{238}U, neglecting a possible dose from cosmic ray exposure:

$$D = (k \cdot D_\alpha + D_\beta + D_\gamma)$$
$$D = (2.36 \cdot 10^{-3} + 1.58 \cdot 10^{-3} + 1.24 \cdot 10^{-3}) \text{ [Gy/y]}$$
$$D = 5.18 \cdot 10^{-3} \text{ [Gy/y]} \qquad (5.12)$$
$$\text{Age} = \frac{\text{paleodose}}{\text{annual dose}} = \frac{8.86 \text{ [Gy]}}{5.18 \cdot 10^{-3} \text{ [Gy/y]}} = 1710 \text{ [y]}.$$

This value can then be used to calculate the age of the pottery sample (Figure 5.21).

TL dating is a compelling and inexpensive method. TL dating on pottery can reach back to earliest pottery samples, which are circa 10,000 years ago. The age limit for TL dating mostly depends on the kind of mineral being used for making the artifact and its quartz content. The amount of natural radioactivity in the mineral determines establishing the annual exposure. In addition, there are saturation effects that are dependent on the crystal structure of the minerals, which limit the number of TL trapping possibilities. By filling all of the trap configurations, a flattening of the growth curve may occur around 10,000–15,000 years. Typical saturation occurs at a total accumulated dose of 100–500 Gy. The saturation level is different for different materials,

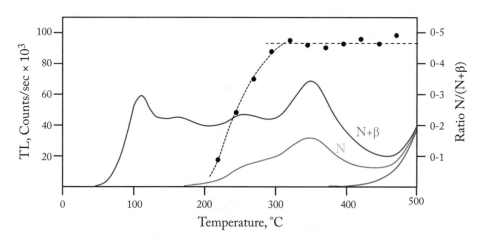

Figure 5.21: Visualization of the plateau method, showing the emitted amounts of thermoluminescence from a piece of pottery as function of temperature for a sample and the sample exposed to an additional dose of 10 Gray of β radiation, yielding a plateau ratio of $R = 0.47$.

depending on the mineral content and annual dose, therefore, corrections for these saturation effects need to be applied!

Another major application for TL is the dating of flint stone tools and weaponry produced by prehistoric people. Flint stone has a longer age range than clay-based material, such pottery or porcelain. Flint is the name given to chalcedony nodules that occur in chalk. Chalcedony is a form of quartz and TL characteristics similar to quartz inclusions in pottery clay are expected. Flint was used as major tool material in Paleolithic times and is therefore ideally suited for thermoluminescence dating.

There are two handicaps with the application of TL for flint; first, the external γ-radioactivity dominates the annual dose, which is therefore substantially lower than for pottery material. Second, middle, and lower Paleolithic flint is often not sufficiently burned, potentially causing errors due to possible pre-Paleolithic TL that has not been reset by a firing process.

To remove TL that may have accumulated from external exposure, outer layers have to be removed in order to only analyze the inner flint sections, which have not been exposed to external α or β radiation from the soil material, so as to assure that the annual dose is only based on the internal γ radiation. After removing outer material, the flint stone cannot be exposed to light, to avoid light induced luminescence effects!

Due to the lack of α and β doses, the γ dose is only 10^{-3} Gy/y. A saturation is reached much later at 100,000–500,000 y. The possibility for reaching 1,000,000 y exists for samples with very low radioactivity. Because of the increased age range, TL dating of flint has been a perfect tool for exploring the early history of humankind in the Paleolithic age (50,000–10,000 BC)

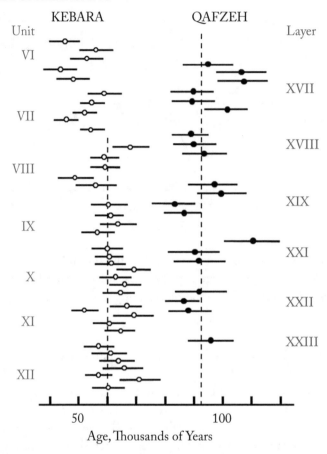

Figure 5.22: The results of TL dating of different flint stone tools at the Kebara and Qafzeh caves in Israel. The layers at which different flint samples were found at the two sites are shown as a function of age. The figure demonstrates that the Kebara site was used by the Neanderthals more than 30,000 years after Qafzeh was populated by Cro-Magnons. Figure plotted from the data presented in [12].

while TL dating of early ceramics has emerged as an important tool for dating the evolution and migration of Neolithic man (10,000–3,000 BC).

A powerful example of the impact of the TL dating of flint stone on our understanding or interpretation of early human history is the analysis of flint tools at different depth layers of the Kebara and Qafzeh caves in Israel [12]. At the Kebara site, a Neanderthal skeleton was discovered and dating of flint stone in the same layer indicated 60,000 y of age. In contrast, at nearby Qafzeh a Cro-Magnon skeleton was discovered with flint stone dating of 92,000 y of age. This is strong evidence of the early presence of Cro-Magnon man (see Figure 5.22).

The great advantage of TL dating is that it requires limited equipment and can be done at a relatively low cost. This makes it the perfect tool for the art market to assess age and value of prehistoric samples or to identify forgeries. The need for these tests is obvious, considering the fact that 40% of all objects tested by the TL laboratories are proven to be reproductions; half the antiquities brought for sale at Sotheby's are imitations; about 5,000 forgeries of ancient art enter the art market each year (https://physicsworld.com/a/antique-dealers-turn-to-physics/).

Nevertheless, the forger community never rests, and it is relatively easy to artificially generate TL in a pottery or porcelain sample, suggesting a much higher age. This is easily done by exposing artifacts to radiation, using an external radiation source. A difficulty, however, is the homogeneous distribution of natural radioactivity versus directed exposure by an external source. This is because efficient exposure can be done only for γ sources; external β exposure would only increase TL from the surface material of the sample, due to the short-range of β radiation. TL tests indeed indicate that several forged porcelain pieces have been artificially β radiated to produce TL signature. A 2-hour irradiation using a ^{40}K source of 10^{8}Bq would convert a 20-year-old piece of modern pottery into a seemingly 1,200-year-old historic artifact. This of course requires access to a strong gamma source. Professional counterfeiters, therefore, insert small pieces of old pottery into forged artifacts at positions that are difficult to see where TL samples are usually taken (https://www.chineseantiques.co.uk/a-quick-tool-for-detecting-forgeries/) (https://archive.archaeology.org/0101/abstracts/africa.html).

5.5   REFERENCES

[1] A. P. Fitzpatrick, The Amesbury archer, *Curr. Archaeol.*, 184:146–152, 2003. DOI: 10.1017/s0003598x00090980. 205

[2] G. D. Farquhar, J. R. Ehleringer, and K. T. Hubick, Carbon isotope discrimination and photosynthesis, *Annu. Rev. Plant Physiol. Plant Mol. Biol.*, 40:503–537, 1989. DOI: 10.1146/annurev.pp.40.060189.002443. 205

[3] N. J. van de Merwe, Carbon isotopes, photosynthesis, and archaeology, *Amer. Scientist*, 70:596–606, 1982. 206

[4] G. Bonani, S. Ivy, I. Hajdas, T. R. Niklaus, and M. Suter, AMS 14C age determinations of tissue, bone, and grass samples from the Ötztal ice man, *Radiocarbon*, 36:247–250, 1994. DOI: 10.1017/s0033822200040534. 208

[5] R. E. M. Hedges, R. A. Housley, C. R. Bronk, and G. J. van Klinken, Radiocarbon dates from the Oxford AMS system: Archaeometry datelist 15, *Archaeometry*, 34:337–357, 1992. DOI: 10.1111/j.1475-4754.1992.tb00507.x. 208

[6] W. Rom, R. Golser, W. Kutschera, A. Priller, P. Steier, and E. M. Wild, AMS 14C dating of equipment from the Iceman and of spruce logs from the prehistoric salt mines of Hallstatt, *Radiocarbon*, 41:183–197, 1999. DOI: 10.1017/s0033822200019536. 208

[7] W. Kutschera and W. Rom, Otzi, the prehistoric Iceman, *Nucl. Instr. Meth. B*, 164:12–22, 2000. DOI: 10.1016/S0168-583X(99)01196-9. 193, 208, 209

[8] P. E. Damon, D. J. Donahue, B. H. Gore, A. L. Hatheway, A. J. T. Jull, T. W. Linick, P. J. Sercel, L. J. Toolin, C. R. Bronk, E. T. Hall, R. E. M. Hedges, R. Housley, I. A. Law, C. Perry, G. Bonani, S. Trumbore, W. Woelfli, J. C. Ambers, S. G. E. Bowman, M. N. Leese, and M. S. Tite, Radiocarbon dating of the Shroud of Turin, *Nature*, 337:611–615, 1989. 208, 211

[9] W. C. McCrone, Judgement day for the Shroud of Turin, *Amherst*, page 288, NY, 1999. 212

[10] T. Casabianca, E. Marinelli, G. Pernagallo, and B. Torrisi, Radiocarbon dating of the Turin Shroud: New evidence from raw data, *Archaeometry*, 61:1223–1231, 2019. DOI: 10.1111/arcm.12467. 212

[11] P. Ball, Is this holy relic preserved? *Nat. Mater.*, 16:503, 2017. DOI: 10.1038/nmat4897. 212

[12] H. Valladas, J. L. Joron, G. Valladas, B. Arensburg, O. Bar-Yosef, A. Belfer-Cohen, P. Goldberg, H. Laville, L. Meignen, Y. Rak, E. Tchernov, A. M. Tillier, and B. Vandermeersch, Thermoluminescence dates for the Neanderthal burial site at Kebara in Israel, *Nature*, 330:159–160, 1987. DOI: 10.1038/330159a0. 220

CHAPTER 6

Summary and Outlook

The discipline called art history is little more than a vast and undefined complex of auxiliary sciences. Each art historian draws the elements of his argument from information provided by all sorts of disciplines, some of them having little to do with art history.

(Roger H. Marijnissen, Paintings, 1985)

Nuclear and atomic physics are certainly among these disciplines, which at first seem to have very little to do with art history or archaeology, yet, over the last decades, have emerged as powerful techniques amidst the traditional approaches in the liberal arts. The power of analyzing the microstructure of matter to determine its composition and origin, and the radioactive impurities of matter, to provide a clock for determining its age, is a unique and rapidly emerging tool.

This book is an attempt to provide an overview of the rich tools that the techniques of nuclear and atomic physics provide for analyzing cultural heritage artifacts. We have by no means given a complete picture of all the analytical tools that have been developed over the last few decades. However, we have attempted in the first section to describe the underlying physical principles these tools are based on, while in the later sections we have addressed specific applications that use spectroscopic- or radiation-based methods.

The methods presented here are mostly the standard approaches, although there are multitudes of possible variations in approach for the testing of material composition to the finest level. The atomic and molecular (vibrational) spectroscopy tools allow endless possibilities for the identification of the composition of materials in artworks and artifacts. The most common methods, including XRF, PIXE, and Raman spectroscopy, enable non-invasive characterization of different objects at different length-scales. Analytical works done with these methods are often complemented by micro-destructive spectroscopic and electron microscopy-based imaging methods that allow in-depth characterization of materials.

Another particular field of research is the understanding of degradation and corrosion patterns of historical materials and artworks. This field often requires performing model experiments of the materials, in artificial and accelerated aging conditions, in order to mimic the processes that happened over the centuries and millennia. Such investigation requires in-depth knowledge in chemistry, materials science, and the underlying physical principles of spectroscopic and imaging tools that are used in this exciting field of research.

During the last decade, an enormous amount of work has been done to not only adopt existing spectroscopic and imaging tools for cultural heritage investigation but also to develop new analytical devices specifically for this field. Interestingly, some of these new devices in the portable or tabletop configurations have been successfully commercialized to satisfy the needs in museum settings and archaeological sites. The current trend is that some of the spectroscopic methods are being combined to create hybrid tools that could simultaneously analyze the composition and molecular makeup of materials. The use of radiography and tomography methods in conjunction with high-resolution electron microscopy enables the investigation of materials in unprecedented detail, on macro, micro, nano, and atomic levels.

There are also a multitude of alternative techniques in the dating of artifacts. In this book, we only addressed the more common one, the radiocarbon method that, next to dendrochronology and thermoluminescence, is able to analyze data from the last 20 millennia of human history. It is, therefore, a perfect tool for tracking human artifacts. However, for going back into prehistoric times, one has to utilize different techniques in order to map anthropological or even geological developments. These approaches include chemical methods such as the FUN test—where the age of a specimen is determined by the ratio of fluorine (F), uranium (U), and nitrogen (N) content that have diffused at different rates into the sample materials. Physics methods are also used—such as isotope analysis, comparing the oxygen isotope ratios in specimens with the isotope ratios in Arctic or Antarctic ice cores of well-defined age, or analyzing the changes in the magnetic alignment of geological layers caused by the switching of the earth's magnetic field, which allows for tracking geological changes over millions of years. Radioactive dating methods—based on longer-lived radioactive species than ^{14}C—are the K-Ar method, which relies on the decay of the ^{40}K radioactive isotope with a half-life of $T_{1/2} = 1.251 \cdot 10^9$ y, or the U-Th dating method, which is based on the analysis of the natural decay chains of ^{238}U ($T_{1/2} = 4.5 \cdot 10^9$ y) and ^{232}Th ($T_{1/2} = 1.405 \times 10^{10}$) emanating from the natural impurity content in the materials. These methods allow us to determine the age of artifacts to more than a million years, if not further. However, a more detailed description of these methods is beyond the scope of this book, which focuses on the analysis of historical artifacts.

The techniques outlined in this book are primarily used to investigate and trace the age and origins of cultural heritage material and also to verify provenance, aiming for a deeper understanding of the historical and cultural environment in which the item was created. This is the research work of the liberal arts, performed at academic institutions, museums, and other art centers. The funding level in the United States is relatively low, particularly as compared with the investment that other countries make in science-based cultural heritage research. Europe, with a multitude of research centers focused on cultural heritage studies, certainly takes the lead in the development and application of these techniques.

A second purpose for the here described applications is to verify the authenticity of specific art samples. While the eye and experience of the art historian is invaluable, scientific study of the detailed composition of the sample adds equally invaluable complementary information. The demand for authenticity is not only driven by the cultural quest underlying art history but is also driven by the economy of the art market. The rising value of paintings and other artifacts finds its response in the increasing number of forgeries, typically announced as "newly found" works of famous painters—from Vermeer to Matisse, from Van Gogh to Jackson Pollock—often explained against the background of the dark history of the 20th century. The forgeries that attract the wealthy customer are not only of the art of the famous but are also of the indigenous art that has been discovered on the international markets of tourism and trade, whether a piece of "ancient" pottery from African markets or a "real" mummy, claimed to be from the looted treasures of the Middle East.

Forgers keep up with the demands of the market and with the improvement in forensic techniques; they develop an increasing sophistication—at least the good ones do—in passing even the most basic of scientific tests, using original, reprocessed materials such as wood, paint, or even radiation exposure to generate artificial TL, in order to pass the more standard thermoluminescence tests. The internet-based trade of historical artifacts may cause an exponential increase in trade and opportunities for experienced forgers.

Scientific analysis is necessary to curb this development but often seems too cumbersome and expensive to keep up with the market. While famous museums like the Louvre have their own institute for analyzing, validating, and restoring their artwork, many museums and art dealers lack these opportunities. In addition, there is a certain level of reluctance on the part of the owners, be it museums or private citizens, to subject their artwork to scientific scrutiny, fearing results that would cause significant devaluation and embarrassment. This issue is beyond the scope of this book; we presented an overview of microscopic research technologies that can address a wide variety of questions in cultural heritage research and the identification of forgeries.

Authors' Biographies

MICHAEL WIESCHER

Michael Wiescher is the Freimann Professor of Physics and the Director of the Nuclear Science Laboratory at the University of Notre Dame. He received his Ph.D. at the University of Münster in 1980. After several years as postdoc and lecturer at Ohio State University, the University of Mainz, Germany, and Caltech, he accepted in 1986 a faculty position at the University of Notre Dame, where he developed a program in nuclear astrophysics using stable and radioactive beams. Dr. Wiescher's research interests are in low energy nuclear physics, with focus on nuclear astrophysics and nuclear applications. The analysis of objects of cultural heritage is such an application of nuclear science. His research is being pursued mainly at the Notre Dame Nuclear Science Laboratory, the CASPAR deep underground accelerator at the Sanford Underground Research Facility in South Dakota, and at several other national and international research institutions. He has published about 400 scientific and review papers and has served on the organizing and advisory committees of nearly 80 national and international conferences. Dr. Wiescher is a Fellow of the American Physical Society and the American Association of the Advancement of Science. In 2003, he was awarded the Hans Bethe Prize in Nuclear Physics and Astrophysics of the American Physical Society and in 2018 he received the Laboratory Astrophysics Award of the American Astronomical Society.

KHACHATUR MANUKYAN

Khachatur Manukyan received his Ph.D. in Chemistry from Yerevan State University (YSU, Armenia) in 2006. He remained at YSU for several years as a lecturer and was also visiting scientist at the Institute of Chemical Physics, National Academy of Sciences (Armenia), before moving to the Department of Chemical and Biomolecular Engineering, University of Notre Dame. In 2013, he joined the Nuclear Science Laboratory at the Department of Physics of the University of Notre Dame as a Research Associate and is currently Research Assistant Professor. Dr. Manukyan's research centers on investigating the different aspects of creating nanoscale materials for applications ranging from energy storage to nuclear astrophysics. Current research topics include materials for nuclear physics and energy applications. He is also actively investigating historical materials such as ancient alloys, cellulose fibers, and pigments in medieval manuscripts. He has published over 80 research articles, invited review papers, and holds several patents in these areas. He has organized professional meetings and currently serves as an editor of the *International Journal of Self-Propagating High-Temperature Synthesis*. Dr. Manukyan is a recipient of the Presidential Prize and Gold Medal of Republic of Armenia in the Field of Natural Sciences (2009), a Fulbright Scholarship Award (2010), and a Best Electron Microscopy Publication Award of Notre Dame Integrated Imaging Facility (2013).

Printed in the United States
by Baker & Taylor Publisher Services